T0234686

SpringerBriefs in Computer Science

SpringerBriefs present concise summaries of cutting-edge research and practical applications across a wide spectrum of fields. Featuring compact volumes of 50 to 125 pages, the series covers a range of content from professional to academic.

Typical topics might include:

- A timely report of state-of-the art analytical techniques
- A bridge between new research results, as published in journal articles, and a contextual literature review
- A snapshot of a hot or emerging topic
- An in-depth case study or clinical example
- A presentation of core concepts that students must understand in order to make independent contributions

Briefs allow authors to present their ideas and readers to absorb them with minimal time investment. Briefs will be published as part of Springer's eBook collection, with millions of users worldwide. In addition, Briefs will be available for individual print and electronic purchase. Briefs are characterized by fast, global electronic dissemination, standard publishing contracts, easy-to-use manuscript preparation and formatting guidelines, and expedited production schedules. We aim for publication 8–12 weeks after acceptance. Both solicited and unsolicited manuscripts are considered for publication in this series.

More information about this series at http://www.springer.com/series/10028

Akka Zemmari • Jenny Benois-Pineau

Deep Learning in Mining of Visual Content

 Springer

Akka Zemmari
Laboratoire Bordelais de Recherche
en Informatique (LaBRI)
University of Bordeaux
Talence Cedex, France

Jenny Benois-Pineau
Laboratoire Bordelais de Recherche
en Informatique (LaBRI)
University of Bordeaux
Talence Cedex, France

ISSN 2191-5768   ISSN 2191-5776   (electronic)
SpringerBriefs in Computer Science
ISBN 978-3-030-34375-0   ISBN 978-3-030-34376-7   (eBook)
https://doi.org/10.1007/978-3-030-34376-7

This Springer imprint is published by the registered company Springer Nature Switzerland AG.
The registered company address is: Gewerbestrasse 11, 6330 Cham, Switzerland

Having written this book for young researchers and professionals, we dedicate it to those who encouraged us, as young researchers, to
Dominique BARBA and Yves METIVIER

Preface

Deep learning is a new trend in artificial intelligence pushing the performances of machine learning algorithms to their real-life applicability for solving a variety of problems the humanity faces nowadays.

Computer vision and multimedia indexing research have realized a general move from all previous approaches to those on the basis of deep learning.

This approach is built on the principles of supervised machine learning and in particular, artificial neural networks. Applied to visual information mining, it also incorporates our fundamental knowledge in image processing and analysis. Today, we translate all our know-how in visual information mining into this language.

The proliferation of software frameworks allows for easy design and implementation of deep architectures, for the choice and adequate parameterization of different optimization algorithms for training parameters of deep neural networks. The availability of graphical processing units (GPU) and of distributed computing made the computational times for learning quite reasonable. For young researchers and those who move to this kind of methods it is important, we think, to get very quickly into comprehension of underlying mathematical models and formalism, but also to make a bridge between the methods previously used for understanding of images and videos and these winning tools.

It is difficult today to write a book about deep learning, so numerous are different tutorials easily accessible on the Internet. What is the particularity of our book compared to them? We tried to keep a sufficient balance between the usage of mathematical formalism, graphical illustrations, and real-world examples. The book should be easy to understand for young researchers and professionals with engineering and computer science background. Deep learning without pain, this is our goal.

Acknowledgements We thank our PhD students, Abraham Montoya Obeso and Karim Aderghal, for their kind help in preparation of illustrative material for this book.

Bordeaux, France   Akka Zemmari
Talence, France   Jenny Benois-Pineau
August 2019

Contents

Acronyms

ANN   Artificial Neural Network
CNN   Convolutional Neural Network
HMM   Hidden Markov Model
LSTM   Long Short-Term Memory Network
MLP   Multi-Layered Perceptron
RNN   Recurrent Neural Network

List of Figures

Chapter 1
Introduction

Visual content mining has a long history and has been a central problem in the field of Computer Vision. It consists in finding and correctly labelling objects in images or video sequences, recognition of static and dynamic scenes. It is necessary in a large set of research and application domains: multimedia indexing and retrieval, computer vision, robotics, computer-aided diagnosis using medical images...Humans are naturally good at performing visual scene recognition without any particular effort. However, automatic object and scene recognition still remains a challenging task.

If we have a look to the history of mathematical models used in Visual content mining, then we can enumerate quite a lot of approaches from a very large field of Pattern Recognition [Tou74]:

- *Statistical pattern recognition*, where a visual scene, its element or even a pixel in fine-grain visual analysis tasks are considered as a point in N-dimensional data space and are seen as realizations of multi-dimensional stochastic process [Tou74];
- *Correlation analysis or template matching [Pra91]*, where the problem consists in recognition of a prototype object in a given visual scene and solved by maximizing correlation function between the prototype and selected areas in the target image or video sequence either in Fourier domain or pixel domain;
- *Structural or syntactic pattern recognition*[Pav77, Fu82], where visual scenes are represented as graphs [Pav77] or sentences from a formal language [Fu82] and the recognition processes consisted in computation of graph isomorphisms or parsing of sentences accordingly to defined formal grammars...

Since the early ages of visual content mining and as in a statistical pattern recognition perspective, there was always the trend to deploy classical data analytics and machine learning for understanding visual content. Hence supervised learning approaches, such as linear discriminant analysis (LDA) [Tou74] or neural network (NN) classifiers, [Min87] have been used.

© The Author(s), under exclusive license to Springer Nature Switzerland AG 2020
A. Zemmari, J. Benois-Pineau, *Deep Learning in Mining of Visual Content*,
SpringerBriefs in Computer Science, https://doi.org/10.1007/978-3-030-34376-7_1

The real break-through in the sense of application of these methods for visual content mining tasks has been achieved with introduction of the Support Vector Machine classifiers (SVMs) by V.N. Vapnik [BV92]. These classifiers outperformed LDA and NNs, which by this time were limited by high-dimensionality of problems to solve requiring training of a significant amount of synaptic weights with a pixel-wise image representation.

In case of visual content classification tasks, the input to the SVMs constituted descriptive vectors extracted from images. Probably, the most simple form of them were the one-dimensional vectors obtained from pixels retinas after histogram equalization and grey-level normalization [HHP01]. Then a very large variety of descriptor vectors have been proposed, such as SIFT [Low04] and SURF [BETVG08], computed around some characteristic points in images. In order to get the global representation of an image and using the achievements of vector quantization, the Bag-of-Visual-Words (BoVW) has been introduced [CDF+04]. The image or an area in it have been presented as a histogram of quantized descriptors and for long years the community was stacked with the combination of BoVW built upon different quantized descriptors for spatial and temporal scenes and SVM classifiers. A comprehensive survey of these methods can be found in [BPPC12]. The main research trends were in how to define the support for descriptor computation and how to build cascade of classifiers fusing visual information encoded as BoVWs on different features [IBPQ14]. For static images, the most popular descriptors were SIFT and SURF expressing contour orientations in the vicinity of characteristic points. Hence the recognition process was based on very local image description abstracted in a global histogram. Some improvements have been obtained by incorporating visual attention models into feature selection and weighting of quantized features in the BoVW histogram [GDBBP16, BPC17], but still these methods got saturated in performance and were not able to achieve descent scores to be used for real-world scene and object recognition.

In 2013, amazing progress was made in object recognition with the presentation of the so-called "region based convolutional neural networks" (R-CNNs) [GDDM13]. Here the convolutional neural networks, an extension of classical multilayer perceptron [Ros61] to the image understanding tasks [LBBH98] were applied to a diverse real-world visual scenes. The method combines selective search [UvdSGS13] and convolutional neural networks classifiers into one algorithm to solve the object recognition task. Improved versions of this algorithm appeared in 2015 [Gir15, RHGS15]. Since then, we can say that for visual content mining there exists one winner approach nowadays: Deep Learning. "Deep" means that compared to limited number of layers in the perceptron, these networks comprise large amount of layers allowing better abstraction from pixels to classes in visual classification tasks.

If we now make a retrospective from Artificial Intelligence (AI) point of view, then we can see that from already the early days of AI as an academic discipline, researchers thought about the possibility for machines to learn from data. Models like perceptrons, which can be seen as simple prototypes of neural networks, were already known, along with probabilistic models.

The beginning of the 80s was widely dominated by expert systems that belong to the logical approach, and statistics and neural networks research were out of favor at the time. The outburst for neural network theory came along with the reinvention of the back-propagation algorithm in the mid 80s and its successful adaptation to neural networks [RHW86]. Since then, Machine Learning and more specifically neural networks have become more and more popular and efficient. In the recent years, a new trend in Machine Learning appeared: researchers started to develop deeper and deeper neural networks, and outstanding results were obtained in many Computer Science problems. In particular, many progress were made in the field of Computer Vision.

Machine Learning (ML), which covers Deep Learning approaches is a branch of Artificial Intelligence (AI) that proposes the use of statistical and algorithmic approaches to exploit large volumes of data. This combination of the two fields AI and Image Analysis and Pattern Recognition turns out to be very successful and rare is research today targeting visual mining tasks which does not use these models.

This book aims at introducing deep learning for visual content mining. In the era of a very intensive research in this field it would be too ambitious to pretend to cover all its aspects and directions. From the perspective of Image Analysis and Pattern Recognition what we are doing now is translating, into "Deep Learning" language, all what we have known to do before, in order to progress in visual content mining tasks. The book is organized in a manner that the reader could get knowledge of theoretical aspects of Deep Learning and to see how these principles are applied to practical visual content mining tasks. Therefore, some chapters of the book are theoretical and can be considered as general introduction to neural networks and deep learning fundamentals. Others are focused on application of deep learning to visual mining tasks.

Chapter 2
Supervised Learning Problem Formulation

In machine learning we distinguish various approaches between two extreme ones: unsupervised and supervised learning. The task of unsupervised learning consists in grouping similar data points in the description space thus inducing a structure on it. Then the data model can be expressed in terms of space partition. Probably, the most popular of such grouping algorithms in visual content mining is the K-means approach introduced by MacQueen as early as in 1967 [Mac67], at least this is the approach which was used for the very popular Bag-of-Visual Words model we have mentioned in Chap. 1. The Deep learning approach is a part of the family of supervised learning methods designed both for classification and regression. In this very short chapter we will focus on the formal definition of supervised learning approach, but also on fundamentals of evaluation of classification algorithms as the evaluation metrics will be used further in the book.

2.1 Supervised Learning

Supervised learning is related to the task of learning a function that maps a set of training samples to known labels. If we consider that the training dataset is composed of the pairs $\{(\mathbf{x}_1, y_1), \ldots, (\mathbf{x}_n, y_n), \ldots, (\mathbf{x}_N, y_N)\}$, where $\mathbf{x}_n \in \mathbb{R}^k = X$ are feature vectors and $y_n \in Y$ the labels of classes, we can consider the function $g(\mathbf{x}, \alpha)$ to map inputs to outputs, such as $X \rightarrow Y$. Then, we can predict labels on unseen data with $g(\mathbf{x}_n, \alpha) = \hat{y}_n$. "The problem of learning is that of choosing from the given set of functions $g(\mathbf{x}, \alpha)$, $\alpha \in \Lambda$, with Λ a set of parameters, the one that best approximates the supervisor's response" [Vap95], i.e. the responses corresponding to known y_n.

Let us now consider a loss function $\mathcal{L}(y_n, \hat{y}_n)$ to measure the error of predicting $\hat{y}_n$. Theoretically we need to minimize the loss function on all the data from our

© The Author(s), under exclusive license to Springer Nature Switzerland AG 2020
A. Zemmari, J. Benois-Pineau, *Deep Learning in Mining of Visual Content*,
SpringerBriefs in Computer Science, https://doi.org/10.1007/978-3-030-34376-7_2

space, but only the training dataset is available, hence we are speaking about
empirical risk minimization:

$$R_{emp}(g) = \frac{1}{N} \sum_{n=1}^{N} \mathcal{L}(y_n, g(\mathbf{x}_n, \alpha)). \tag{2.1}$$

As the result of learning process with known class of functions $g(.)$ we will get
the set of optimal parameters α^*.

With the empirical risk minimization principle covered, is also very important to
minimize the Structural Risk by introducing the penalty $C(g)$ in order to balance
the model's complexity against its success while fitting training pairs [Vap92],

$$J(g) = R_{emp}(g) + \lambda C(g), \lambda \geq 0. \tag{2.2}$$

For a given form of g, such as in SVMs [CV95], K-NNs [CH$^+$67] or MLPs
[Ros61], the problem consists in finding optimal parameters α^*. Where α^* may be
a multidimensional array of parameters for different models to train and $\lambda > 0$
is the "regularization" parameter. Greater we fix it, more we pay attention to the
complexity of the model. An example of structural risk for neural networks would
be

$$E(w, \lambda) = \frac{1}{N} \sum_{n=1}^{N} \mathcal{L}(y_n, g(\mathbf{x}_n, \mathbf{w})) + \lambda \|\mathbf{w}\|^2. \tag{2.3}$$

with $\mathbf{w}$—synaptic weights.

2.2 Classification and Regression

Classification and regression are categorized under the same shadow of supervised
learning approach. Both are based on the same, they use labeled datasets to fit a
function and map inputs to outputs. Then, when we fit a function $g(\mathbf{x}, \alpha^*)$ to get $\hat{y}$,
we seek for the better approximation in order to predict on new data. Here is where
the main question arises between classification and regression, should the output be
continuous or discrete? That is the key to identify if our problem is a classification
task or regression; classification outputs are discrete while regression outputs are
continuous. In Fig. 2.1, we show the main differences between classification and
regression based on provided training data.

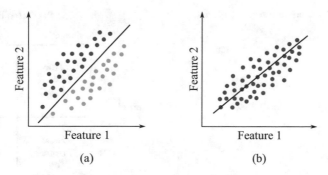

Fig. 2.1 Classification and regression examples given two features. In (**a**), a classification problem where red dots are class 0 and green dots are class 1, the black line represents the model to classify new data into two categories. In (**b**), a regression example where red dots are data and the black line represents the model to predict continuous values

2.3 Evaluation Metrics

To evaluate any classification algorithm, we are fist speaking about the classification errors and then are thinking about time constraints for training and generalization steps. We will further remind the fundamentals in evaluation of classification approaches.

2.3.1 Confusion Matrix

Let us define **P** as the number of positive samples in a benchmark dataset with known labels and **N** the number of negative samples in the dataset. The elements in a confusion matrix represent the errors while classifying positive and negative samples, and allow assessment of the performance of a classification algorithm. As shown in Fig. 2.2, in this matrix, each column represents the instances of predicted classes by a trained classifier while each row represents the instances in an actual class. Thus in a binary classification problem we have a 2×2 confusion matrix where we have the following notations

- True positive: a correct prediction (TP).
- True negative: a correct rejection (TN).
- False positive: a false alarm (FP).
- False negative: a misclassification (FN).

Fig. 2.2 A 2 × 2 confusion
matrix

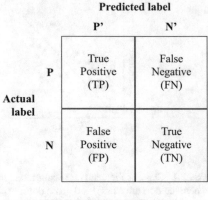

		Predicted label	
		P'	**N'**
Actual label	**P**	True Positive (TP)	False Negative (FN)
	N	False Positive (FP)	True Negative (TN)

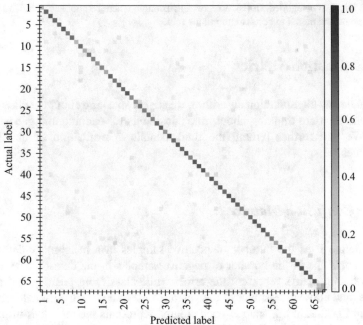

Fig. 2.3 A 67 × 67 confusion matrix example in the task of architectural classification, see Chap. 8. Values are normalized in the range [0, 1]

For the case of multi-class classification, false negatives between classes are easily identified using this matrix. In the main diagonal, accuracies per class are concentrated, in each row, confusion level is displayed based on miss-classified samples. An example of a confusion matrix is presented in Fig. 2.3 in the task of architectural classification in 67 categories.

2.3.2 Metrics

After training a model, different evaluation methods are applied to measure how good the results are. Based on the nature of the tasks in computer vision, we must select a suitable evaluation metric that describes the performance of trained models allowing also the comparison with other methods. For example, different metrics are used in the tasks of: image classification, object localization, segmentation or image retrieval. In the following we will present some of them related to classification tasks.

Classification accuracy is the term we usually use when we speak about accuracy in machine learning. This corresponds to the number of correct predictions out of the total number of input samples. It is commonly used when datasets are balanced, with the same number of samples per class. The accuracy (ACC) is computed as,

$$ACC = \frac{(TP + TN)}{(TP + FP + TN + FN)}.$$

(2.4)

If the testing dataset is not balanced, a better option is to use Balanced Accuracy (BACC) to better interpret classification results. It is defined as the average of recall for each class that is $\frac{TP}{P}$ for positive class and $\frac{TN}{N}$ for negative class. The BACC metrics is in particular used when evaluating performance of classification algorithms in medical image analysis[BABPA$^+$17] as the classes with pathology, e.g. Alzheimer Disease (see Chap. 9) are less populated than normal control subjects in cohorts.

$$BACC = \frac{\frac{TP}{P} + \frac{TN}{N}}{2}.$$

(2.5)

In the case of multi-class classification, such as in [KSH12] for ImageNet dataset with 1000 categories, the *accuracy at top-k* is used to measure the proportion of samples correctly classified when the true label is in the top-k predictions (the k ones with the highest probabilities).

True Positive Rate (TPR), also called *sensitivity* or *recall*, measures the proportion of actual positives samples that are correctly identified as positive samples,

$$TPR = \frac{TP}{(TP + FN)},$$

(2.6)

while True Negative Rate (TNR), also called *specificity*, measures the proportion of actual negative samples that are correctly identified as negative samples,

$$TNR = \frac{TN}{(TN + FP)}.$$

(2.7)

We also remind the *precision*:

$$P = \frac{TP}{(TP + FP)}.$$

(2.8)

We note that *recall* (R) and *precision* (P) are the metrics which have been widely used in visual information retrieval[BPPC12].

Here, in the case of unbalanced classes of images the F-score metrics is also used.

$$F = \frac{2}{\frac{1}{R} + \frac{1}{P}}.$$

(2.9)

The False Positive Rate (FPR) or *fal-out*, is calculated as the ratio between the number of negative samples wrongly categorized as positive (FP) and the total number of actual negative samples, measuring the false alarm ratio,

$$FPR = \frac{FP}{(TN + FP)}.$$

(2.10)

Finally, the False Negative Rate (FNR) or *miss-rate*, measures the portion of miss-classified samples.

$$FNR = \frac{FN}{(FN + TP)} = 1 - TPR.$$

(2.11)

2.3.3 AUC-ROC Curve

If we consider a binary classification problem which depends on one or more thresholds or parameters for discrimination, the Area Under Curve of the Receiver Operating Characteristic (AUC-ROC curve) is a plot to illustrate and measure how the classifier performs when a parameter is varied. The ROC curve is a probability curve, while AUC tells how capable the model is to distinguish between different classes [Faw06].

The ROC curve is plotted with TPR (Eq. (2.6)) against FPR (Eq. (2.11)), on y-axis and x-axis, respectively. In this plot, TPR defines how many correct positive predictions occur among all positive samples available during the test. FPR, on the other hand, defines how many incorrect positive results occur among all negative samples available during the test.

As FPR and TPR are in the range [0, 1], both are computed at different threshold values between [0, 1] in order to draw the graph in Fig. 2.4. Then, is evident that when the AUC is high (close to 1 in the range [0, 1]) the model is better to discriminate between categories.

Fig. 2.4 The AUC-ROC
curve plot

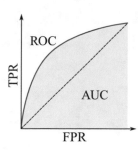

In a multi-class classification model, we can plot N number of AUC-ROC Curves for C classes using "one" vs "all" methodology. For example, if we have three categories labeled as 0, 1 and 2, first, we plot 0 classified against 1 and 2, then we plot class 1 classified against 0 and 2 and finally, class 2 against 0 and 1. For a general perspective, we can average the values in the curves.

2.4 Conclusion

Hence in this chapter we have introduced the general principles of supervised learning Deep learning being a part of it. We further recalled definitions of metrics allowing for evaluation of performances of these algorithms. In the follow up of the book we will focus on the Deep learning starting from the fundamental model of neural network classifiers.

Chapter 3
Neural Networks from Scratch

Artificial neural networks consist of distributed information processing units. In this chapter, we define the components of such networks. We will first introduce the elementary unit: the formal neuron proposed by McCulloch and Pitts in [McC43]. Further we will explain how such units can be assembled to design simple neural networks.

Then we will discuss how a neural network can be trained for classification.

3.1 Formal Neuron

The neuron is the elementary unit of the neural network. It receives input signals $(x_1, x_2, \cdots, x_p)$, applies an activation function f to a linear combination z of the signals. This combination is determined by a vector of weights $(w_1, w_2, \cdots, w_p)$ and a bias b. More formally, the output neuron value y is defined as follows:

$$y = f(z) = f\left(\sum_{i=1}^{p} w_i x_i + b \right).$$

Figure 3.1a sums up the definition of a formal neuron. A compact representation of the same is given in Fig. 3.1b.

Different activation functions are commonly encountered in neural networks:
The step function ξ_c:

$$\xi_c(x) = \begin{cases} 1 & \text{if } x > c \\ 0 & \text{otherwise.} \end{cases} \tag{3.1}$$

This simplistic function was the first activation function considered. Its main problem is that it can activate different labels to 1, and the problem of classification

© The Author(s), under exclusive license to Springer Nature Switzerland AG 2020
A. Zemmari, J. Benois-Pineau, *Deep Learning in Mining of Visual Content*,
SpringerBriefs in Computer Science, https://doi.org/10.1007/978-3-030-34376-7_3

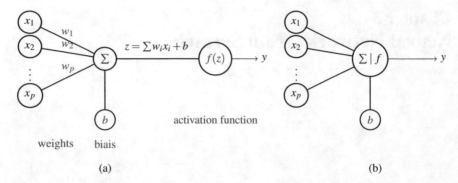

Fig. 3.1 Two representations of a formal neuron: a detailed representation (**a**) and a compact one (**b**). Note that when it is a part of a neural network, the neuron is simply represented by a vertex

is not solved. As a consequence, the use of smooth functions is preferred, as it gives analog activations rather than binary ones, thus the risk of having several labels scored 1 is widely reduced for smooth activation functions.

The sigmoid function σ:

$$\sigma(x) = \frac{1}{1 + e^{-x}}. \tag{3.2}$$

It is one of the most popular activation functions. It maps $\mathbb{R}$ onto the interval $[0, 1]$. This function is a smooth approximation of the step function, see Fig. 3.2a. It has many interesting properties.

The continuity of the function enables to properly train networks for non-binary classification tasks. Its differentiability is a good property in theory because of the way neural networks "learn". Furthermore, it has a steep gradient around 0, which means that this function has a tendency to bring the y values to either end of the curve: this is a good behaviour for classification, as it makes clear distinctions between predictions. Another good property of this activation function is that it is bounded: this prevents divergence of the activations.

The biggest problem of the sigmoid is that it has very small gradient when the argument is distant from 0. This is responsible for a phenomenon called the vanishing of the gradient: the learning is drastically slowed down, if not stopped.

The tanh function:

$$\tanh(x) = \frac{2}{1 + e^{-2x}} - 1 = 2\sigma(2x) - 1. \tag{3.3}$$

As the above equation suggests, this function is in fact a scaled and vertically shifted sigmoid function that maps $\mathbb{R}$ onto the interval $[-1, 1]$, thus it shares the same pros

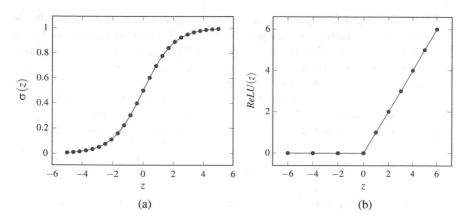

Fig. 3.2 Two particular activation functions: (**a**) the sigmoid function and (**b**) the Rectified Linear Unit (Relu) function

and cons. The main difference between tanh and sigmoid lies in the strength of the gradient: tanh tends to have stronger gradient values. Just like the sigmoid, tanh is also a very popular activation function.

The ReLU function:

$$\text{ReLU}(x) = \max(0, x). \tag{3.4}$$

The Rectified Linear Unit (ReLU) (Fig. 3.2b) has become very popular in the recent years. It indeed has very interesting properties:

- This function is a simple thresholding of the activations. This operation is simpler than the expensive exponential computation in sigmoid and tanh activation functions.
- ReLU tends to accelerate training a lot. It supposedly comes from its linear and non-bounded component.
- Unlike in sigmoid-like activations where each neuron fires up in an analog way which is costly, the 0-horizontal component of ReLU leads to a sparsity of the activations, which is computationally efficient.

It also has its own drawbacks:

- Its linear component is non-bounded, which may lead to an exploding activation.
- The sparsity of activations can become detrimental to the training of a network: when a neuron activates in the 0-horizontal component of ReLU, the gradient vanishes and training stops for this neuron. The neuron "dies". ReLU can potentially make a substantial part of the network passive.
- It is non-differentiable in 0: problem with the computation of the gradient near 0.

Many other activation functions exist. They all have in common the non-linear property, which is essential: if they were linear, the whole neural network would be linear (a linear combination is fed to a linear activation, which is fed to a linear combination etc.), but in this case no matter how many layers we have, those linear layers would be equivalent to a unique linear layer and we would then loose the multi-layer architecture characteristic to neural networks: the final layer would simply become a linear transformation applied to the input of the first layer.

A particular activation function is also of high interest: the Softmax function. We will discuss it later.

3.2   Artificial Neural Networks and Deep Neural Networks

An artificial Neural Network (ANN) incorporates the biological geometry and behaviour into its architecture in a simplified way: a neural network consists of a set of layers disposed linearly, and each layer is a set of (artificial) neurons. The model is simplified so that signals can only circulate from the first layer to the last one.

Each neuron in the ith layer of the neural network is connected to all the neurons in the $(i - 1)^{th}$ layer, and the neurons in a given layer are all independent from each other. Figure 3.3 gives two examples of simple artificial neural networks. It is "feedforward network" as the information flows from the input **x** to the output **y** only in one direction. It is also important to note that *all* the neurons of the same layer have the *same* activation function.

A network with all the inputs connected directly to the outputs is a single-layer network. Figure 3.3a shows an example of such a network. In [McC43], McCulloch and Pitts prove that a network with a single unit can represent the basic Boolean functions AND and OR. However, it is also easy to prove that a network with a single layer cannot be used to represent XOR function.

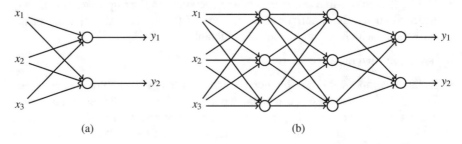

(a)   (b)

Fig. 3.3 Two neural networks: both have three inputs x_1, x_2 and x_3, and two outputs y_1 and y_2. Network (**a**) is a single-layer neural network and (**b**) a multilayers network with two hidden layers

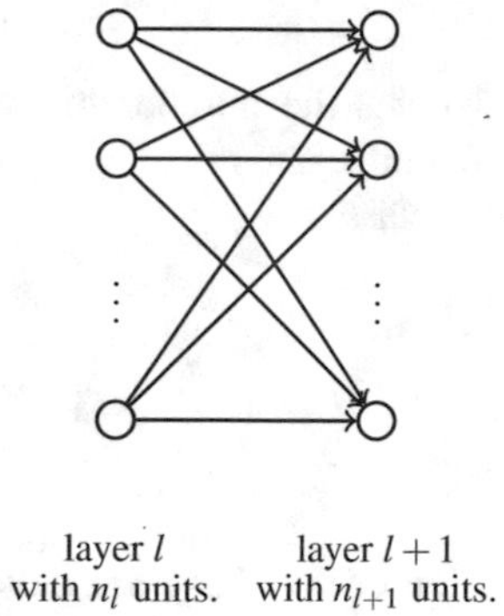

$$w^{(l)} = \begin{pmatrix} w_{1,1}^{(l)} & w_{1,2}^{(l)} & \cdots & w_{1,n_{l+1}}^{(l)} \\ w_{2,1}^{(l)} & w_{2,2}^{(l)} & \cdots & w_{2,n_{l+1}}^{(l)} \\ \vdots & \vdots & \vdots & \vdots \\ w_{n_l,1}^{(l)} & w_{n_l,2}^{(l)} & \cdots & w_{n_l,n_{l+1}}^{(l)} \end{pmatrix} \quad \mathbf{b}^{(l)} = \begin{pmatrix} b_1^{(l)} \\ b_2^{(l)} \\ \vdots \\ b_{n_i}^{(l)} \end{pmatrix}$$

layer l layer $l+1$
with n_l units.   with n_{l+1} units.

Fig. 3.4 Two layers, the weights matrix between the two layers l and $l+1$ and the bias vector of layer l

All the neurons of the same layer l have the same activation function $f^{(l)}$. That is, given a neural network with $L > 0$ layers, if we denote by $\mathbf{y}^{(l)}$ the vector with the outputs of layer l, for any $1 < l < L$, then, one can write:

$$\mathbf{y}^{(l+1)} = f^{(l+1)} \left(w^{(l)^T} \mathbf{y}^{(l)} + \mathbf{b}^{(l+1)} \right) \tag{3.5}$$

Expression (3.5) explains how the values given as input to the network are forwarded to compute the value of the output $\hat{\mathbf{y}} = \mathbf{y}^{(l+1)}$ (Fig. 3.4).

Any neural network must contain at least two layers: one for input and one for output. As discussed, some functions cannot be implemented using a neural network without any hidden layer (XOR gate for example). However, Cybenko, in [Cyb89], proved that any continuous function can be approximated with any desired accuracy, in terms of the uniform norm, with a network of one hidden layer whose activation functions are sigmoids. A *deep* neural network is an artificial neural network with *at least one* hidden layer.

3.3 Learning a Neural Network

A neural network is a supervised machine learning model (see Chap. 2). It learns a prediction function from a training set [Vap92]. Each sample from this set can be modeled by a vector which describes the observation and its corresponding response. The learning model aims to construct a function which can be used to predict the responses for new observations while committing a prediction error as lowest as possible.

The training dataset is a pair $(\mathbf{x}, \mathbf{y})$ with:

- $\mathbf{x} = (x_{i,j})_{1 \leq i \leq N, 1 \leq j \leq p}$ a matrix, where N denotes the number of available examples in the dataset and p is the number of features, and
- $\mathbf{y} = (y_i)_{1 \leq i \leq N}$ a vector whose entries are the true classes.

3.3.1 Loss Function

There are many functions used to measure prediction errors. They are called *loss functions*. A loss function somehow quantifies the deviation of the output of the model from the correct response. We are speaking here about "empirical loss"Ì functions [Vap92], that is the error computed on all available ground truth training data.

A neural network is a supervised machine learning model. That is it can be seen as a function f which for any input x computes a predicted value $\hat{\mathbf{y}}$:

$$\hat{\mathbf{y}} = f(x).$$

3.3.2 One-Hot Encoding

Back to the training set, the known response of each observation is encoded in a *one-hot labels* vector. More formally, let $\mathscr{C} = \{c_1, c_2, \cdots, c_k\}$ be the set of all possible classes. That is, given an observation $(\mathbf{x}, y) = (x_1, x_2, \cdots, x_p, y)$, we have $y \in \mathscr{C}$.

We introduce a binary vector $\mathbf{v} = (v_1, v_2, \cdots, v_k)$ such that $v_j = 1$ if $y = c_j$ and $v_i = 0$ for all $i \neq j$. This is the function which ensures a "hard" encoding of class labels. In the sequel, $\mathbf{v}$ will represent the one-hot encoding of the class. Figure 3.5 gives an example of a one hot encoding.

Fig. 3.5 Example of a one-hot encoding

$$\mathscr{C} = \{'Cat','Dog','Horse','Bird'\}$$

$$y = 'Cat' \Rightarrow \mathbf{v} = (1,0,0,0)$$

$$y = 'Horse' \Rightarrow \mathbf{v} = (0,0,1,0)$$

3.3.3 Softmax or How to Transform Outputs into Probabilities

Given a vector $\mathbf{y} = (y_1, y_2, \cdots, y_k)$ with positive real-valued coordinates, the softmax function aims to transform the values of $\mathbf{y}$ to a vector $\mathbf{s} = (p_1, p_2 \cdots, p_k)$ of real values in the range $(0, 1)$ that sum to 1. More precisely, it is defined for each $i \in \{1, 2, \cdots, k\}$ by:

$$p_i = \frac{e^{y_i}}{\sum_{j=1}^{k} e^{y_j}}.$$

The softmax function is used in the last layer of multi-layer neural networks which are trained under a cross-entropy (we will define this function in next paragraphs) regime. When used for image recognition, the softmax computes the estimated probabilities, for each input data, of being in an image class from a given taxonomy.

3.3.4 Cross-Entropy

The *cross-entropy* loss function is expressed in terms of the result of the softmax and the one-hot encoding. Recall that our neural network takes as input an example $\mathbf{x}$ from the training dataset, then it computes the probabilities:

$$\hat{\mathbf{y}} = (\hat{y}_1, \hat{y}_2, \cdots, \hat{y}_k),$$

where $\hat{y}_i$ is the probability for the class of example $\mathbf{x}$ to be in c_i. Recall also that v is the one-hot encoding of the ground truth y of example $\mathbf{x}$.

The cross-entropy is defined as follows:

$$D(\hat{\mathbf{y}}, y) = -\sum_{i=1}^{k} v_i \ln(\hat{y}_i). \tag{3.6}$$

The definition of one-hot encoding and Eq. (3.6) mean that only the output of the classifier corresponding to the correct class label is included in the cost.

The aim of the training phase is to build a model (the neural network and is related to maximum likelihood as we will explain in next paragraphs) that can compute the probabilities for each example in the training dataset to be in a given class. Thus if M is the trained model, if it is given an example having class c_i, then the probability assigned by M to this example will be $\hat{y}_i$. Hence, if the dataset contains n_i examples from class c_i, then the probability that all these examples will be correctly classified by the model M is $\hat{y}_i^{n_i}$. Now, if we denote by $C = \{c_1, c_2, \cdots, c_k\}$ the set of classes of the whole examples, then the probability that the model M gives the right class to each example is given by:

$$\mathbb{P}r\left(C \mid M\right) = \prod_{i=1}^{k} \hat{y}_i^{n_i}. \tag{3.7}$$

Equality (3.7) can be rewritten as follows:

$$\ln\left(\mathbb{P}r\left(C \mid M\right)\right) = \sum_{i=1}^{k} n_i \ln\left(\hat{y}_i\right). \tag{3.8}$$

Now, if we divide (3.8) by the size N of the training dataset we obtain:

$$\frac{1}{N} \ln\left(\mathbb{P}r\left(C \mid M\right)\right) = \frac{1}{N} \sum_{i=1}^{k} n_i \ln\left(\hat{y}_i\right). \tag{3.9}$$

Since we are using one-hot encoding, one can see that $\frac{n_i}{N}$ is simply the ground truth.

The goal of the training phase is to find a model M which maximizes the probability in expression (3.7). If we take the negatives in expression (3.9), we get the definition of the *Average Cross-Entropy*:

$$\mathcal{L}(y, \hat{y}) = \frac{1}{N} \sum_{i=1}^{N} D(\hat{y}^{(i)}, y^{(i)}), \tag{3.10}$$

where $y^{(i)}$ (respectively $\hat{y}^{(i)}$) is the ground-truth (respectively the probability computed by the network) for the ith example.

3.4   Conclusion

In this chapter we introduced the elements of neural networks. We defined the elementary unit in such networks: the formal neuron. Then we discussed deep networks: networks with many hidden layers. This is the simplest architecture for deep learning. In next chapters, we will discuss more powerful architectures which are specialized for image recognition and dynamic visual contents.

We also defined the loss function which measures the error between the ground truth and the predicted classes. In the next chapter, we will present the optimization methods used to find the parameters of the network.

Chapter 4
Optimization Methods

The machine learning models aim to construct a prediction function which minimizes the loss function. There are many algorithms which aim to minimize the loss function. Most of them are iterative and operate by decreasing the loss function following a descent direction. These methods solve the problem when the loss function is supposed to be convex. The main idea can be expressed simply as follows: starting from initial arbitrary (or randomly) chosen point in the parameter space, they allow the "descent" to the minimum of the loss function accordingly to the chosen set of directions. Here we discuss some of the most known and used optimization algorithms in this field.

The chapter presents the foundations of optimization methods. We use different notations from previous chapter but we will show how the methods can be applied for neural networks learning.

4.1 Gradient Descent

The use of gradient based algorithms has proven to be very effective in order to optimize the numerous parameters of neural networks, thus it is one of the most common approaches for network learning. As a result, many state-of-the-art Deep Learning libraries contain extensive implementation of the different forms of gradient algorithms. We briefly recall facts concerning the gradient descent algorithm.

Let us denote by θ the vector containing all the parameters of the neural network, and let $J(\theta, g(\mathbf{x}), y)$ be the cost function which represents the error between the ground-truth labels $g(\mathbf{x})$ (associated with the training data $\mathbf{x}$) and the predicted data y estimated by the neural network under parameters θ. If one imagines the cost function representation as a valley, the idea behind gradient algorithm is to follow the slope of the mountain until we reach the bottom of the surface. Indeed, the

© The Author(s), under exclusive license to Springer Nature Switzerland AG 2020
A. Zemmari, J. Benois-Pineau, *Deep Learning in Mining of Visual Content*,
SpringerBriefs in Computer Science, https://doi.org/10.1007/978-3-030-34376-7_4

direction given by the slope is the one which decreases the most locally, and it is actually the opposite to gradient. A thorough mathematical discussion about the gradient descent can be found in [WF12].

The iteration step for the gradient descent is given by:

$$\boldsymbol{\theta}_{t+1} \leftarrow \boldsymbol{\theta}_t - \eta \nabla_{\boldsymbol{\theta}} J \left(\boldsymbol{\theta}, g(\mathbf{x}), y \right), \tag{4.1}$$

where η is a non negative constant called learning rate. This method works in spaces of any dimension (even infinite) and can be used for both linear and non-linear functions. To be well-posed, the gradient descent requires the function to be L-Lipschitz, that is

$$\forall \boldsymbol{u}, \boldsymbol{v} \in \mathbb{R}^N, \ \|\nabla J(\boldsymbol{u}) - \nabla J(\boldsymbol{v})\|_2 \le L \|\boldsymbol{u} - \boldsymbol{v}\|_2, \tag{4.2}$$

where L is Lipschitz constant. Moreover, the algorithm is only guaranteed to converge to the global solution in the case where the function J is strictly convex. If that's not the case, the algorithm will not even be guaranteed to find a local minimizer.

In the case of neural network learning, especially in deep learning, one wishes to optimize a huge number of parameters, that is the cost function arguments will be in a very high dimensional space, which leads to a proliferation of saddle points, see Fig. 4.1, known to be potentially harmful when looking for a minimizer, since they tend to be surrounded by high error plateaus [DPG+14], see Fig. 4.2.

Besides, choosing properly the learning rate is crucial for the method to be efficient: it has to be small enough to enable the convergence of the algorithm, though it should not be too small as it would dramatically slow down the process. Carefully and iteratively selecting this step size thus may help improving the results. A classical algorithm used for this purpose is the line search algorithm [Hau07].

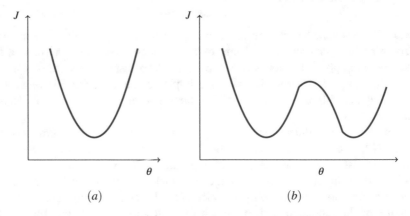

$$(a) \qquad\qquad\qquad (b)$$

Fig. 4.1 Two functions: (**a**) a convex function having a single minimum and (**b**) a non-convex one having two local minima

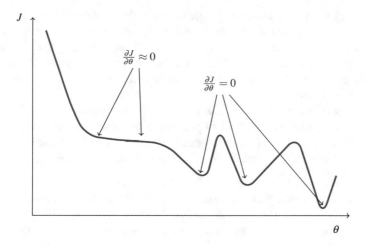

Fig. 4.2 A non-convex function with multiple local minima and plateaus

4.2 Stochastic Gradient Descent

The problem encountered with gradient descent algorithms in the context of Deep Learning is that the mathematical properties required for the problem to be well-posed are not met in general. Indeed, in such very high dimensional context, the loss function can be non-convex and non-smooth, thus the convergence property is not satisfied, and convergence toward local minima can be extremely slow.

An excellent way to address those problems is to use the stochastic gradient descent algorithm. This algorithm is a stochastic approximation of the gradient descent algorithm. It aims at minimizing an objective function which can be written as a sum of differentiable functions (typically in the context of image processing, one function per image). Such process is done iteratively over batches of data (that is subsets of the whole dataset) randomly selected. Each objective function minimized this way approximates the "global" objective function. The following formula resumes this method:

$$\boldsymbol{\theta}_{t+1} \leftarrow \boldsymbol{\theta}_t - \eta \frac{1}{B_s} \sum_{i=1}^{B_s} \nabla_\theta J_t^i(\boldsymbol{\theta}_t, \mathbf{x}_t^i, y_t^i), \tag{4.3}$$

where B_s is the batch size, $B_t = (\mathbf{x}_t^i, y_t^i)_{i \in [|1, B_s|]}$ is the batch of data associated with step t, where the $\mathbf{x}_t^i$ and the y_t^i are respectively the ground truth data and the estimated data in this batch, and $J_t = \frac{1}{B_s} \sum_{i=1}^{B_s} J_t^i$ is the stochastic approximation of the global cost function at step t over the batch B_t, decomposed into a sum of differentiable functions J_t^i associated to each pair $(\mathbf{x}_t^i, y_t^i)$.

Stochastic gradient descent (SGD) addresses most of the concerns encountered with gradient descent in the context of Deep Learning:

- The stochasticity of the method helps with the weak mathematical setting: although the problem is ill-posed with a high dimensional objective function that is non-convex and potentially non-smooth, stochasticity tends to improve convergence as it helps the objective function to pass through local minima and saddle points, which are known to be surrounded by very flat plateaus in high dimensional spaces.
- Convergence is much faster, as it is way less costly to perform many update steps over small batches of data than to perform a single update step over the whole dataset (in the context of Deep Learning, and specifically in visual data mining, datasets can contain several thousands of images or video frames).

This comes with its own drawbacks: at first, stochastic approximation of the gradient descent means that convergence toward the global minimizer cannot be guaranteed. Moreover, the smallest are the batches used for SGD, and the more variance can be observed in the results. As a consequence, many architectures are designed to take advantage of both sides, by selecting an average value of batch size.

4.3   Momentum Based SGD

The main purpose of momentum based approaches is to accelerate the gradient descent process. This is done by expanding the classical model with a velocity vector, which will build up iterations after iterations. A physical perspective is that such gradient acceleration mimics the increase in kinetic energy while a sphere is rolling down a valley.

In terms of optimization behaviour, it appears that (stochastic) gradient descent struggles to descend into a region where the surface of objective function curves more steeply in one direction than in another.

The iteration step for the momentum variant of the gradient descent is given below:

$$
\begin{aligned}
v_{t+1} &\leftarrow \mu v_t - \eta \nabla J(\theta_t), \\
\theta_{t+1} &\leftarrow v_{t+1} + \theta_t.
\end{aligned}
\tag{4.4}
$$

The vector v_{t+1} (initialized at zero) is computed first at each iteration and represents the update of the velocity of the "sphere rolling down the valley". The velocity stacks at each iteration, hence the need of the hyper-parameter μ, in order to damp the velocity when reaching a flat surface, otherwise the sphere would move too much near local extrema. A good strategy is to change the value of μ depending on learning stage.

In the velocity update, two terms are competing: the accumulated velocity μv_t and the negative gradient at the current point. The key idea is that in the scenario described in the previous paragraphs (when the surface curves more steeply in a direction than in another), then the two terms of the velocity will not have the same direction. This will prevent the gradient term from oscillating too much and thus will speed up the convergence. The importance of momentum and initialization in Deep Learning was discussed in a recent paper (2013) [SMDH13a] of Ilya Sutskever et al.

4.4 Nesterov Accelerated Gradient Descent

In 1983, Nesterov proposed in [Nes83] a slight modification of the "classical momentum" and showed that his algorithm had an improved theoretical convergence for convex function optimization. This approach has become very popular, as it usually performs better in practice than the classical momentum, and is still a good fit for gradient optimization even today.

The key difference between the Nesterov and the classical momentum algorithm is that the latter computes first the gradient at the current location θ_t and then performs a step in the direction of the accumulated velocity, whereas the Nesterov momentum first performs the step, which gives an approximation of the updated parameter which we call $\widetilde{\theta}_{t+1}$, and corrects this step by computing the gradient at this new location.

To understand the reasoning behind this difference, one can understand the gradient term $-\eta \nabla J(\theta_t)$ in the update v_{t+1} of the velocity as a correction term of the accumulated velocity μv_t. It makes more sense to correct an error (here the step performed by the accumulated velocity) after the error has been made, that is to compute the gradient at the location $\widetilde{\theta}_{t+1}$. The iterative step for the Nesterov momentum is:

$$
\begin{aligned}
\widetilde{\theta}_{t+1} &\leftarrow \theta_t + \mu v_t, \\
v_{t+1} &\leftarrow \mu v_t - \eta \nabla J(\widetilde{\theta}_{t+1}), \\
\theta_{t+1} &\leftarrow v_{t+1} + \theta_t.
\end{aligned}
\tag{4.5}
$$

4.5 Adaptative Learning Rate

We saw that in the stochastic gradient descent algorithm, the learning rate is defined as a non-negative constant. This can be a source of error: when the gradients are small but consistent (near local extrema for instance), a strong learning rate leads to oscillations in the valley, which prevents the method from converging properly. What we want is to move slowly in directions with strong but inconsistent gradients, and conversely to move quickly in directions with small but consistent gradients (see Fig. 4.3).

One can combine these two properties by adapting the learning rate adaptively. Some common annealing schedules used for this purpose are detailed below:

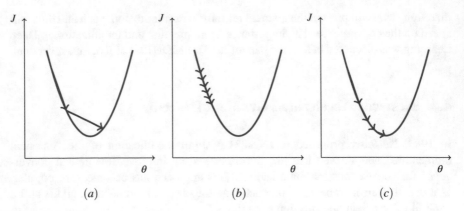

Fig. 4.3 Three examples of learning rates: (**a**) a large rate can yield to a divergent behaviour, (**b**) a too small rate yield to a slow convergence and (**c**) an adaptive rate can ensure a quick convergence

- Step decay: After each k epochs, multiply the learning rate by a constant $C < 1$.
- Polynomial decay: Set the learning rate as

$$\forall t \geq 0, \ \eta_t = \frac{a_0}{1 + b_0 t^n}, \quad a_0, b_0 \in \mathbb{R}_+. \tag{4.6}$$

- Exponential decay: Set the learning rate as

$$\forall t \geq 0, \ \eta_t = a_0 e^{-b_0 t}, \quad a_0, b_0 \in \mathbb{R}_+. \tag{4.7}$$

These strategies have the drawback to be arbitrary, and might not be suited for a specific learning problem. A thorough discussion regarding proper learning rate tuning for SGD methods was conducted by LeCun in [SZL13]. An other possibility is to propose an adaptive learning rate from the computation of the inverse of the Hessian matrix of the cost function (Newton and quasi-Newton approaches). The corresponding iterative step is

$$\widetilde{\boldsymbol{\theta}}_{t+1} \leftarrow \boldsymbol{\theta}_t - [HJ(\boldsymbol{\theta}_t)]^{-1} \nabla J(\boldsymbol{\theta}_t). \tag{4.8}$$

The latter method is an example of second order optimization procedure. More information regarding Newton's optimization method can be found in [Chu14].

In practice, it can be costly to compute the second order derivatives and to perform matrix inversion. Furthermore, stability problems in Hessian Matrix inversion can be encountered near local extrema. Therefore, this is not a well suited method in the context of Deep Learning.

4.6 Extensions of Gradient Descent

Here we recall briefly some of the existing expansions of the gradient descent:

Averaged Gradient Descent This particular version of the gradient descent, studied in the paper of Polyak [PBJ92] replaces the computation of the parameter values θ_t by the computation of the temporal mean values, from the updates obtained through gradient descent:

$$\bar{\theta}_T \leftarrow \frac{1}{T} \sum_{t=0}^{T} \theta_t. \tag{4.9}$$

Adagrad In this method first introduced in a 2011 paper [DHS11], the goal was to have the learning rate adjust itself, depending on the sparsity of the parameters. In this context, the sparsity of a parameter means that this parameter has not been trained a lot during the iterative training process. Sparse parameters will learn faster while non-sparse ones will learn slower. This emphasizes the training of sparse features that would not have been trained properly in a classical gradient descent algorithm. The corresponding update step is different for each parameter $(\theta)_i$ in the parameter vector θ. It is given by

$$\forall i, \ (\theta_{t+1})_i \leftarrow (\theta_t)_i - \alpha \frac{(\nabla J(\theta_t))_i}{\sqrt{\sum_{u=1}^{t} (\nabla J(\theta_u))_i^2}}, \quad \alpha > 0. \tag{4.10}$$

RMSProp Just like in the case of Adagrad, the RMSProp algorithm proposes to adjust automatically the learning rate of each parameter. It does so by running average of the magnitudes of recent gradients for that parameter. This algorithm was presented in the course [HSS12]. The corresponding update step is given by the formula below:

$$\forall i, \ (\nabla_{t+1})_i \leftarrow \delta(\nabla_t)_i + (1 - \delta)\left(\nabla J(\theta_t)\right)_i^2,$$
$$\forall i, \ (\theta_{t+1})_i \leftarrow (\theta_t)_i - \alpha \frac{(\nabla J(\theta_t))_i}{\sqrt{(\nabla_{t+1})_i}}, \quad \alpha > 0. \tag{4.11}$$

The parameter δ sets the confidence given either to the running average of the magnitudes of past gradients or to the magnitude of the last gradient computed.

Adam Adam algorithm is among the most recent and efficient first-order gradient descent based optimization algorithms. It was first presented in [KB14]. As in the case of Adagrad and RMSProp, it automatically adjusts the learning rate for each of the parameters. The particularity of this method is that it computes so called "adaptative moment estimations" (m_t, v_t). This method can be seen as a generalization of the Adagrad algorithm. Its corresponding update process is detailed below:

$$\forall i, \ (m_{t+1})_i \leftarrow \beta_1 \cdot (m_t)_i + (1 - \beta_1) \cdot \big(\nabla J(\theta_t)\big)_i,$$
$$\forall i, \ (v_{t+1})_i \leftarrow \beta_2 \cdot (v_t)_i + (1 - \beta_2) \cdot \big(\nabla J(\theta_t)\big)_i^2,$$
$$\forall i, \ (\theta_{t+1})_i \leftarrow (\theta_t)_i - \alpha \frac{\sqrt{1-\beta_2}}{1-\beta_1} \frac{(m_t)_i}{\sqrt{(v_t)_i}+\epsilon}, \qquad \alpha, \epsilon > 0 \text{ and } \beta_1, \beta_2 \in]0, 1[.$$

$$(4.12)$$

In the previous equation, ϵ is used as a precision parameter. The parameters β_1 and β_2 are used to perform running averages over the so-called moments m_t and v_t respectively.

4.7 Gradient Estimation in Neural Networks

The previous sections introduce general optimization methods. In this section, we consider this methods used for neural networks optimization. That is the cost function J is the loss function $\mathcal{L}$, the parameters θ correspond to the parameters of the neural network:

$$\theta = \left(w^{(l)}, b^{(l+1)}\right), \quad \text{for } l \in \{0, 1, \cdots, L - 1\}.$$

One of the main reasons for the resurgence of neural networks was the development of an algorithm which efficiently computes the gradient of the cost function, that is all the partial derivatives with respect to weights and biases in the neural network.

The key idea behind this algorithm, in the case of feed-forward neural networks, is that a slight modification of the weights and the bias in a layer l has a (slight) impact on the next layer, which cascades up to the output layer. Thus, in order to compute the partial derivatives with respect to weights and biases, we focus on the analysis of the errors (the slight modifications of the outputs), and we do it in a backward manner (hence the name backpropagation), since the loss function of our neural network depends directly on the activations in the output layer. In other words, we try to understand how the errors in the output layer propagate iteratively from a layer to the previous layer.

In next sections we explain the backpropagation algorithm. We mainly explain how to compute the derivatives of the loss function with respect to the weights and the bias. It is obvious that the update of the parameters is done using one of the expressions we explained before but the reader can simply apply Eq. (4.1) once the derivatives are computed.

4.7.1 A Simple Example

Let first consider a very simple neural network: a network with a single hidden layer with only two neurons, one in each layer (see Fig. 4.4). We also consider that the

Fig. 4.4 A very simple neural network: two layers with a single unit in each layer. The two neurons have the same activation function σ

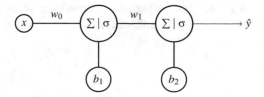

activation function for the two neurons is the sigmoid function ($\sigma(z) = \frac{1}{1+e^{-z}}$) and we use a simple definition of the loss function:

$$\mathcal{L} = \frac{1}{2}(\hat{y} - y)^2.$$

For the sake of simplicity, we also use the following notation: $w^{(i)} = (w_i)$ and $b^{(i+1)} = (b_{i+1})$ for $i \in \{0, 1\}$.

Thus, the expression of $\hat{y}$ is simply:

$$\hat{y} = \sigma(z_2) \tag{4.13}$$

where

$$z_2 = w_1 y_1 + b_2, \; y_1 = \sigma(z_1) \text{ and } z_1 = w_0 x + b_1.$$

Recall that, if we take as learning rate the value $\eta = 1$, the gradient descent step in this case becomes, for all $t > 1$ and for $i \in \{0, 1\}$:

$$w_i^{(t+1)} = w_i^{(t)} - \frac{\partial \mathcal{L}}{\partial w_i}\left(w_0^{(t)}, w_1^{(t)}\right)$$
$$b_{i+1}^{(t+1)} = b_{i+1}^{(t)} - \frac{\partial \mathcal{L}}{\partial b_{i+1}}\left(b_1^{(t)}, b_2^{(t)}\right)$$

The key point is how to compute the four partial derivatives. For this, we use the *chain rule*. Thus for $u \in \{w_1, b_2\}$, one can write:

$$\frac{\partial \mathcal{L}}{\partial u} = \frac{\partial \mathcal{L}}{\partial \hat{y}} \times \frac{\partial \hat{y}}{\partial z_2} \times \frac{\partial z_2}{\partial u}.$$

Computing each derivative, we get:

$$\frac{\partial \mathcal{L}}{\partial \hat{y}} = \hat{y} - y, \quad \frac{\partial \hat{y}}{\partial z_2} = \hat{y}(1 - \hat{y}), \quad \frac{\partial z_2}{\partial w_1} = y_1 \text{ and } \frac{\partial z_2}{\partial b_2} = 1,$$

yielding:

$$\frac{\partial \mathcal{L}}{\partial w_1} = (\hat{y} - y)\hat{y}(1 - \hat{y})y_1 \text{ and } \frac{\partial \mathcal{L}}{\partial b_2} = (\hat{y} - y)\hat{y}(1 - \hat{y}).$$

To compute the partial derivatives according to w_0 and b_1, we use again the chain rule but with a slightly different form:

$$\frac{\partial \mathcal{L}}{\partial u} = \frac{\partial \mathcal{L}}{\partial y_1} \times \frac{\partial y_1}{\partial z_1} \times \frac{\partial z_1}{\partial u},$$

where $z_1 = w_0 x + b_1$.

On the other hand:

$$\frac{\partial y_1}{\partial z_1} = y_1(1 - y_1), \quad \frac{\partial z_1}{\partial w_0} = x, \text{ and } \frac{\partial z_1}{\partial b_1} = 1,$$

and, using again the chain rule:

$$\frac{\partial \mathcal{L}}{\partial y_1} = \frac{\partial \mathcal{L}}{\partial z_2} \times \frac{\partial z_2}{\partial y_1}$$

and from the fact that:

$$\frac{\partial \mathcal{L}}{\partial z_2} = \frac{\partial \mathcal{L}}{\partial \hat{y}} \times \frac{\partial \hat{y}}{\partial z_2} = (\hat{y} - y)\hat{y}(1 - \hat{y}),$$

we obtain:

$$\frac{\partial \mathcal{L}}{\partial w_0} = (\hat{y} - y)\hat{y}(1 - \hat{y})y_1(1 - y_1)w_1 x \text{ and } \frac{\partial \mathcal{L}}{\partial b_1} = (\hat{y} - y)\hat{y}(1 - \hat{y})y_1(1 - y_1)w_1.$$

4.7.2   General Case: Backpropagation Algorithm

The backpropagation algorithm uses gradient descent to find the parameters that minimize the loss function. It is a recursive method based on chain rule. In the sequel, we consider a neural network with hidden layers and sigmoid as activation function for all the neurons (including the output layer). We also consider that the loss function is the cross-entropy function.

We use the same notation as in Fig. 3.3. Thus, the output of the network is a vector $\hat{\mathbf{y}} = (\hat{y}_1, \hat{y}_2, \cdots, \hat{y}_{n_L})$ whose elements are given by:

$$\hat{y}_i = y_i^{(L)} = \sigma\left(z_i^{(L)}\right)$$

$$= \sigma\left(\sum_{k=1}^{n_{L-1}} w_{k,i}^{(L-1)} y_k^{(L-1)} + b_i^{(L)}\right). \tag{4.14}$$

The cross-entropy for a single example is given by the sum:

$$\mathcal{L} = -\sum_{i=1}^{n_L} \left(y_i \ln(\hat{y}_i) + (1 - y_i) \ln(1 - \hat{y}_i)\right).$$

To use the gradient descent iteration (see Eq. (4.1)), we need to compute the derivatives of the loss function accordingly to parameters. We first consider the parameters in the output layer (layer L).

Let $w = w_{k,i}^{(L-1)}$ be a weight between a unit in layer $L - 1$ and a unit i in the output layer. We need to compute the value of $\frac{\partial \mathcal{L}}{\partial w}$. For this, we use the chain rule:

$$\frac{\partial \mathcal{L}}{\partial w} = \frac{\partial \mathcal{L}}{\partial \hat{y}_i} \times \frac{\partial \hat{y}_i}{\partial z_i^{(L)}} \times \frac{\partial z_i^{(L)}}{\partial w}.$$

We also have:

$$\frac{\partial \mathcal{L}}{\partial \hat{y}_i} = -\frac{y_i}{\hat{y}_i} + \frac{1 - y_i}{1 - \hat{y}_i} = \frac{\hat{y}_i - y_i}{\hat{y}_i(1 - \hat{y}_i)}. \tag{4.15}$$

Since

$$\frac{\partial \hat{y}_i}{\partial z_i^{(L)}} = \hat{y}_i(1 - \hat{y}_i), \tag{4.16}$$

and $\frac{\partial z_i^{(L)}}{\partial w} = y_k^{(L-1)}$, we finally obtain:

$$\frac{\partial \mathcal{L}}{\partial w} = \left(\hat{y}_i - y_i\right) \hat{y}_k^{(L-1)}. \tag{4.17}$$

A similar calculation yields to:

$$\frac{\partial \mathcal{L}}{\partial b_i^{(L)}} = \left(\hat{y}_i - y_i\right). \tag{4.18}$$

The above gives the gradients with respect to the parameters in the last layer of the network. Computing the gradients with respect to the parameters in hidden layers requires another application of the chain rule.

Let $w = w_{k,i}^{(l)}$ be a weight between a unit k in a hidden layer l and a unit i in layer $l + 1$ for $l \in \{1, 2, \cdots, L - 2\}$. Then:

$$\frac{\partial \mathcal{L}}{\partial w} = \frac{\partial \mathcal{L}}{\partial y_i^{(l+1)}} \times \frac{\partial y_i^{(l+1)}}{\partial z_i^{(l+1)}} \times \frac{\partial z_i^{(l+1)}}{\partial w}. \tag{4.19}$$

Let $\{u_1, u_2, \cdots, u_m\}$ be the set of units in layer $l + 2$ connected to unit i. We consider $\mathcal{L}$ as a function $\mathcal{L}\left(y_1^{(l+2)}, y_2^{(l+2)}, \cdots, y_m^{(l+2)}\right)$ of the outputs of units u_i in the layer $l + 2$. Then:

$$\frac{\partial \mathcal{L}}{\partial y_i^{(l+1)}} = \sum_{j=1}^{m} \frac{\partial \mathcal{L}}{\partial y_j^{(l+2)}} \times \frac{\partial y_j^{(l+2)}}{\partial y_i^{(l+1)}}$$

$$= \sum_{j=1}^{m} \frac{\partial \mathcal{L}}{\partial y_j^{(l+2)}} \times \frac{\partial y_j^{(l+2)}}{\partial z_j^{(l+2)}} \times \frac{\partial z_j^{(l+2)}}{\partial y_i^{(l+1)}}$$

$$= \sum_{j=1}^{m} \frac{\partial \mathcal{L}}{\partial y_j^{(l+2)}} \times y_j^{(l+2)}(1 - y_j^{(l+2)}) \times w_{i,j}^{(l+1)}. \qquad (4.20)$$

Since $\frac{\partial y_i^{(l+1)}}{\partial z_i^{(l+1)}} = y_i^{(l+1)}(1 - y_i^{(l+1)})$ and $\frac{\partial z_i^{(l+1)}}{\partial w} = y_k^{(l)}$, we can derive the desired derivative:

$$\frac{\partial \mathcal{L}}{\partial w} = y_i^{(l+1)}(1 - y_i^{(l+1)})y_k^{(l)} \times \sum_{j=1}^{m} \frac{\partial \mathcal{L}}{\partial y_j^{(l+2)}} \times y_j^{(l+2)}(1 - y_j^{(l+2)}) \times w_{i,j}^{(l+1)}. \quad (4.21)$$

Equation (4.21) means that the derivative with respect to weights in the hidden layers can be calculated if all the derivatives with respect to the outputs of the next layer are known. This defines a recursive algorithm.

A similar computation can be done to get the derivatives with respect to the bias $b = b_i^{(l+1)}$:

$$\frac{\partial \mathcal{L}}{\partial b} = y_i^{(l+1)}(1 - y_i^{(l+1)})y_k^{(l)} \times \sum_{j=1}^{m} \frac{\partial \mathcal{L}}{\partial y_j^{(l+2)}} \times y_j^{(l+2)}(1 - y_j^{(l+2)}). \quad (4.22)$$

4.8 Conclusion

In this chapter, we presented optimization methods used to train neural networks. In practice, the learning process can take a very long time, especially for deep neural networks, whose associated cost functions are often non-quadratic, non-convex, high-dimensional and with many local minima and valleys (saddle points). In particular, the backpropagation algorithm neither guarantees that the network will converge to a good solution, nor that the convergence will be swift, nor that the convergence will occur at all. In [Sím96], the author showed that training a three-node sigmoid network with an additional constraint on the output activations (such as zero threshold) is NP-hard. Two years later, LeCun investigates in [LBOM98] on a number of tricks that improve the chances to get a satisfying solution while speeding up convergence by possibly several orders of magnitude.

Chapter 5
Deep in the Wild

In this chapter we are interested in how from high-resolution images and videos passing them through a Deep convolutional neural network we get reduced dimension which finally allows a classification decision. We are interested in two operations: convolution and pooling and trace analogy with these operations in a classical Image Processing framework.

5.1 Introduction

With proliferation of image acquisition devices and increasing capacities of storage be it on general-purpose computers, local storage of hand-carried devices, such as mobile phones, touchpads, micro-SD cards integrated in glasses and allowing us to store first-person view of our life, visual information such as images and videos becomes of higher and higher spatial and temporal resolution. Some examples show us the tremendous quantity of pixels an image can contain today captured with a simple mobile phone: it is of 3200×2187 pixels (8 Mpx) and this resolution is not a limit yet. If we consider video, then from 64×64 pixel resolution image on the first mobile-phones with screens in 90s we came to High Definition (HD) video with 1920×1080 pixels up 120 frames per second. Ultra High Definition TV format (UHDTV1) today proposes 3840×2160 pixels per frame and its further version UHDTV2 proposes 7680×4320 pixels per frame. All these formats are being quickly available on all acquisition and visualization supports be it high-resolution TV screens or mobile devices.

The same progression is observed in Digital cinema where from 2K spatial resolution with 2048×1080 pixels per image the digital projection world moved to 4K format which is 4096×2160 ppi and the 8K format is coming with 8192×4320 resolution (source Wikipedia).

A. Zemmari, J. Benois-Pineau, *Deep Learning in Mining of Visual Content*, SpringerBriefs in Computer Science, https://doi.org/10.1007/978-3-030-34376-7_5

To these spatial resolutions adds the colour depth. Today it is unthinkable not to use the full richness of colour information in images and video to perform scene classification and recognition, object detection and recognition, actions recognition in video. Hence the quantity of information has to be tripled when considering pixel representations as vectors in three-dimensional spaces of colour systems.

Visual content mining approaches have to face these tremendous quantity of information per image whatever the source and the target recognition task is: (i) organization, searching and browsing through digital cinema and professional (TV) video archives, or (ii) mining of User Generated Content(UGC) coming from mobile devices and largely present on social networks. Despite classical neural network classifiers such as MLP [Ros58] were successfully used for face recognition in 90s [ByFL99], these detectors worked with very small input images, i.e. small retinas of size 20×20 pixels extracted from images and using 30 neurons in the hidden layer of MLP. These resolutions have nothing to do with the wild amount of pixels, even if we select regions in them, we have today. The solution for the "wild" large-scale problems when classifying visual scenes comprising billions of pixels is in very well known image processing methods: convolution and sub-sampling. Deep neural networks today comprise convolutional and sub-sampling (or, the so called) pooling layers which allow for a drastic dimension reduction of the input high resolution images or video frames. In the follow-up of the chapter we remind convolution and (sub)sampling operations to which we are accustomed to in image processing and analysis and try to make an analogy with the way they are used in convolutional neural networks.

5.2   Convolution

Convolution operation has been known in image processing since the early adventure of filtering algorithms. Let us denote $I(x, y)$ a scalar continuous function of two variables defined on an infinite support $\mathscr{R}^2$, i.e. $I : R^2 \mapsto R$. And let us consider a function $K(x, y)$ similarly defined, we will call it "kernel". Then the convolution operation is expressed by the following equation:

$$\hat{I}(x, y) = (I * K)(x, y) = \iint I(u, v) K(x - u, y - v) \, du \, dv \qquad (5.1)$$

here $\hat{I}$ is the result of convolution. In the discrete case the images and convolutional kernels are defined on a discrete grid of pixels. More than that, the grid of pixels is a finite support for image of size $W \times H$ with W width and H height of the image. The kernel support is usually of smaller size $N \times N$ than the image support $N <<$ W and $N << H$. This ensures only local dependence of the result $\hat{I}(x, y)$ from neighbouring pixels. Thus in a discrete form the convolution operation is expressed by the following equation:

$$\hat{I}(i, j) = (I * K)(i, j) = \sum_{v=-N/2}^{v=N/2} \sum_{\mu=-N/2}^{\mu=N/2} I(\mu, v) K(i - \mu, j - v) \qquad (5.2)$$

For the sake of simplicity we denote continuous functions image and kernel and their discrete versions by the same symbols $I, K, \hat{I}$.

Convolution is a linear operation what can easily be seen from Eq. (5.1). Indeed

$$((\alpha I + b) * K)(x, y) = \iint (\alpha I(u, v) + b) K(x - u, y - v) \, du \, dv \qquad (5.3)$$

$$= \alpha \iint I(u, v) K(x - u, y - v) \, du \, dv+ \qquad (5.4)$$

$$b \iint K(x - u, y - v) \, du \, dv \qquad (5.5)$$

and for normalized kernels we simply have $\alpha(I * K) + b$.

Among convolution kernels used for linear image filtering we distinguish: low-pass, high-pass and pass-band filters. Without going into spectral theory of these filters (we refer the interested reader to fundamental image processing textbooks such as [Pra91]) we just say, that

- low-pass kernels allow for image smoothing;
- high-pass kernels allow for highlighting contrasts and are used for namely contour detection;
- pass-band kernels filter out high-frequency noise and highlight meaningful contrasts such as contours in images.

We illustrate these different kernels effects on an image below. First of all the most popular linear filter is the Gaussian filter. Its kernel expression, in continuous case, is as following

$$K(u, v) = A \exp -\frac{u^2 + v^2}{2\sigma^2} \qquad (5.6)$$

Here A is a normalization constant ensuring that

$$\iint K(u, v) \, du \, dv = 1 \qquad (5.7)$$

or, in a discrete case, that all coefficients of the kernel (5.2) sum to 1 thus ensuring the preservation of the local mean in the image. σ is the scale parameter ensuring stronger or lesser smoothing. In discrete case we are speaking about "kernel mask", i.e. the matrix of coefficients of the kernel. The kernel mask size N depends on the scale parameter and is usually chosen as three times σ. Hence on the border of the kernel $K(\mu, v)$ the kernel values—also called coefficients of the filter are close to

(a)   (b)

Fig. 5.1 Low-pass filtering of an image: (**a**) source image and (**b**) Gaussian filtered image with a 3 × 3 kernel

zero. This relation can be used inversely: for a given size of kernel we can chose the scale parameter. Stronger the size of the kernel is more small details will be filtered in the image. Thus the size of kernel support is always chosen as a function of our problem: if an image is of low resolution or we wish to preserve small details it is useless to chose the filter size more than 3 × 3. The same is hold for the choice of sizes of convolution filters for Convolutional Neural Networks. In Chap. 9 we will speak about shallow-deep networks for brain scan classification and will chose only 3 × 3 kernel sizes to fit low resolution of MRI images of human brain. An example of Gaussian filter application with the size 3 × 3 on a natural image is presented in Fig. 5.1. In the image (b) textures are smoothed and impulse noise is filtered out (low-pass filtering).

An example of convolution with a high-pass filter is given in Fig. 5.2. It is the famous Sobel [Pra13] horizontal kernel which allows to highlight horizontal structures in the image: (a) original image, (b) result of convolution with "horizontal" Sobel mask which expression is given below

$$k(u, v) = \frac{1}{4} \begin{bmatrix} -1 & -2 & -1 \\ 0 & 0 & 0 \\ 1 & 2 & 1 \end{bmatrix}. \tag{5.8}$$

Finally, as an example of pass-band filter we give the Difference-Of-Gaussians (DOG) filter. It is well known in visual information processing as it is the first-step computation of the Scale Invariant Feature Transform (SIFT) key-points proposed by Lowe [Low04].

$$K(u, v) = \frac{1}{2\pi\sigma^2} \exp{-\frac{u^2 + v^2}{2\sigma^2}} - \frac{1}{2\pi \times k^2\sigma^2} \exp{-\frac{u^2 + v^2}{2k^2\sigma^2}} \tag{5.9}$$

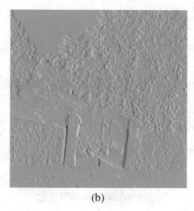

(a)   (b)

Fig. 5.2 (a) Source image and (b) Sobel filtering on x-axis

For the known property of the linearity of convolution, see Eq. (5.3) this filtering means that two results of convolution of original image with two different Gaussian filters:one with the scale parameter σ and another with $k \times \sigma$ are subtracted. The parameter k is chosen as $k > 1$. This filter allows for removing of high-frequency noise on the contours while preserving a good localization of them. An example of images after DOG filtering are presented in Fig. 5.4.

The convolution operation is invariant with regard to translation of image. That is if we consider the translation mapping $\tau_{p,q} I(x, y) = I(x - p, y - q)$, then

$$(\tau_{p,q} I) * K = \tau_{p,q}(I * K) \tag{5.10}$$

This property is well illustrated in Fig. 5.4. Indeed, despite the shift of the stone pillar in the bottom image, the filter response is the same (see the region encircled). Obviously, we do not pretend on precise conservation of pixel values: natural images contain lightening changes and noise.

Now, what happens in any CNN architecture we are using today. The so-called "convolutional layers" perform the convolution operation on input feature maps. The first feature map is the original image. In Fig. 5.3 is illustrated one of the most known architectures AlexNet [KSH12]. The convolution masks superimposed on the image and further feature maps are depicted as black squares. The main difference between the classical convolution filters which are synthesized based on signal processing theory as high-pass, low-pass and pass-band filters with a priori defined coefficients, in convolution masks in CNNs, the coefficients *are trained* when minimizing target objective, i.e. loss function (see Chap. 3). Hence for each dataset and classification task, these filter coefficients should vary. Usually, in the design of Deep architectures, such hyper-parameters as the number and the size of filters for each convolutional layer are chosen empirically, but with the objective to highlight image details with regard to the target classification of regression tasks. In Fig. 5.5 results of application of a convolution filter from the first layer of AlexNet,

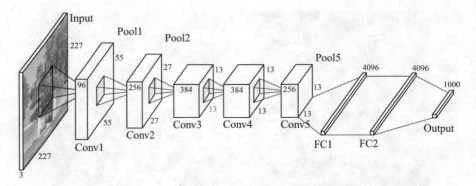

Fig. 5.3 AlexNet architecture [KSH12]

Fig. 5.4 Difference of
Gaussians filtering on two
samples. Samples are shown
in top and bottom rows with
$\sigma = 1.0$ and $\sigma = 1.5$,
respectively

trained in image classification task on ImageNet database [KSH12] and the DOG
filter we previously used (see Fig. 5.4) are presented. As in AlexNet, the convolution
mask is sliding with a step of four pixels in x and y directions, the DOG filtering
of our images was performed with the same shift. This explains the reduced size
of resulting images. When analyzing them we can state that the trained filter acts
similarly to DOG, it highlights contours and smoothes homogeneous areas in the
image (Fig. 5.5).

Fig. 5.5 Feature maps from two samples by applying Difference of Gaussians (DoG) filtering and the output of the first convolutional layer (Conv) of AlexNet architecture [KSH12]

5.3 Sub-sampling

The process of propagation of original high-resolution image trough a Deep CNN comprises three main steps: (i) convolution, (ii) pooling, (iii) non-linear transformation imitating a neuron firing function. Pooling is equivalent to the operation well-known in image processing as "sub-sampling". The latter reduces resolution of the input feature map obtained after the convolution of the original image or feature map from the previous layer of the network. In order to understand what happens in this layer, we will say few words about image sub-sampling and first start with a reminder of sampling principles.

5.3.1 Image Sampling

If we consider our continuous, infinite-extent image $I(x, y)$ (see Eq. (5.1)) and wish to transform it into a digital discrete image $I(i, j)$, then we have to multiply it by a spatial sampling function [Pra91] $S(x, y)$ defined as

$$S(x, y) = \sum_{l=-\infty}^{l=\infty} \sum_{k=-\infty}^{k=\infty} \delta(x - l\Delta x, y - k\Delta y). \tag{5.11}$$

Here $\delta(.)$ is Dirac delta function and thus $S(x, y)$ is a sum of an infinite number of them arranged in a grid with the step sizes Δx and Δy. Our sampled image—we will denote it as $I_P(x, y)$ choosing P for "pixels" is thus the product

$$I_P(x, y) = I(x, y)S(x, y) = \sum_{l=-\infty}^{l=\infty} \sum_{k=-\infty}^{k=\infty} I(l\Delta x, k\Delta y)\delta(x - l\Delta x, y - k\Delta y)$$
$$\tag{5.12}$$

As Pratt writes [Pra91], for the analysis purposes it is convenient to consider spatial frequency domain representation in the continuous Fourier transform domain

$$F_P(\omega_x, \omega_y) = \iint I_P(x, y) \exp\{-i(\omega_x, \omega_y)\} \, dx \, dy \tag{5.13}$$

by the Fourier transform convolution theorem, the Fourier transform of the sampled image $I_P(x, y)$, which is a product of continuous image $I(x, y)$ and sampling function $S(x, y)$, see Eq. (5.12) can be expressed as the convolution of Fourier transforms $F_I(\omega_x, \omega_y)$ and $F_S(\omega_x, \omega_y)$ of $I(x, y)$ and of $S(x, y)$:

$$F_P(\omega_x, \omega_y) = \frac{1}{4\pi^2} F_I(\omega_x, \omega_y) * F_S(\omega_x, \omega_y) \tag{5.14}$$

The Fourier transform of Sampling function $S(x, y)$

$$F_S(\omega_x, \omega_y) = \frac{4\pi^2}{\Delta x \Delta y} \sum_{l=-\infty}^{l=\infty} \sum_{k=-\infty}^{k=\infty} \delta(\omega_x - l\omega_{xs}, \omega_y - k\omega_{ys}) \qquad (5.15)$$

Here $\omega_{xs} = \frac{2\pi}{\Delta x}$ and $\omega_{ys} = \frac{2\pi}{\Delta y}$ are the Fourier domain sampling frequencies. Developing convolution of Eq. (5.14) we get

$$F_P(\omega_x, \omega_y) = \frac{1}{\Delta x \Delta y} \iint F_I(\omega_x - \alpha, \omega_y - \beta)$$
$$\times \sum_{l=-\infty}^{l=\infty} \sum_{k=-\infty}^{k=\infty} \delta(\omega_x - l\omega_{xs}, \omega_y - k\omega_{ys}) \, d\alpha \, d\beta \qquad (5.16)$$

Changing the order of summation and integration and using the shifting property of Dirac delta function, the sampled image spectrum becomes

$$F_P(\omega_x, \omega_y) = \frac{1}{\Delta x \Delta y} \sum_{l=-\infty}^{l=\infty} \sum_{k=-\infty}^{k=\infty} F_I(\omega_x - l\omega_{xs}, \omega_y - k\omega_{ys}) \qquad (5.17)$$

An illustration of this in Fourier domain is given in Fig. 5.6. Figure 5.6a depicts the spectrum of ideal image, $F_I(\omega_x, \omega_y)$, Fig. 5.6b illustrates the spectrum of sampled image. Accordingly to the Eq. (5.17) it consists of the spectrum of ideal image infinitely repeated over the frequency plane on the grid with the step-size $\frac{2\pi}{\Delta x}$, $\frac{2\pi}{\Delta y}$, in horizontal and vertical frequency directions respectively. If the spectrum of the ideal image has a finite support, that is $F_I(\omega_x, \omega_y)$ is defined on the domain $-\omega_{xc} < \omega_x < \omega_{xc}, -\omega_{yc} < \omega_y < \omega_{yc}$ as it is the case illustrated in the figure and the step-size of the grid is sufficiently large, then the reconstruction of ideal continuous image from the sampled image can be seen as a selection/filtering of spatial spectrum of sampled image with a simple box filter first (see illustration in Fig. 5.6c and computation of inverse Fourier transform after that.

The transfer function of a box filter, which is the Fourier transform of its impulse response is expressed as

$$B(\omega_x, \omega_y) = \begin{cases} A & \text{for } |\omega_x| \leq \omega_{xL} \text{ and } |\omega_y| \leq \omega_{yL} & (5.18) \\ 0 & \text{otherwise} & (5.19) \end{cases}$$

Here ω_{xL} and ω_{yL} are frequency extents of the filter support.

Then the reconstruction of the ideal image can be done as

$$I_R(x, y) = F^{-1}(F_P(\omega_x, \omega_y) B(\omega_x, \omega_y)). \qquad (5.20)$$

Fig. 5.6 Sampled image spectra: (**a**) original image spectrum, (**b**) sampled image spectrum and (**c**) reconstructed image spectrum

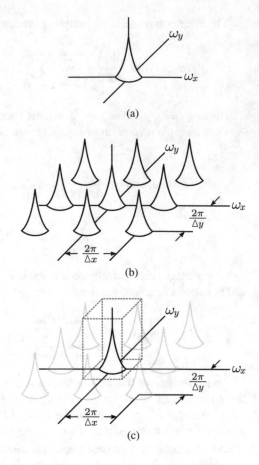

Here F^{-1} denotes the inverse Fourier transform. Despite the Eq. (5.20) is general and concerns not only the box filter, the ideal reconstruction with the latter is only possible if the sampling frequencies $\frac{2\pi}{\Delta x}, \frac{2\pi}{\Delta y}$ are sufficiently high with respect to the cut-off frequencies ω_{xc}, ω_{xc} and satisfy the Nyquist conditions:

$$\omega_{xs} \geq 2\omega_{xc}, \omega_{ys} \geq 2\omega_{yc} \tag{5.21}$$

or equivalently the sampling step-sizes have to satisfy

$$\Delta x \leq \frac{\pi}{\omega_{xc}}, \Delta y \leq \frac{\pi}{\omega_{yc}} \tag{5.22}$$

this is the case illustrated in Fig. 5.6c.

The box filter has two disadvantages. According to the convolution theorem for Fourier transform, the product of spectrum with filter transfer function in spectral domain see Eq. (5.20) equals to the convolution of original image with impulse

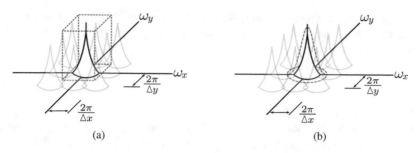

Fig. 5.7 Filtering in Fourier domain: (**a**) with Box filter, (**b**) with Gaussian filter

response of the filter in pixel domain. The impulse response of the Box filter contains $sinc$ functions $\frac{sin(\omega_{xL}x)}{\omega_{xL}x}$ and $\frac{sin(\omega_{yL}y)}{\omega_{yL}y}$ which have infinite support and converges slowly (we send the readers for more details to e.g. [Pra91]). Hence to filter-out spectrum replications in the spectrum of sampled image for ideal reconstruction, the filtering has to be performed in the Fourier domain, which requires direct and then inverse Fourier transform computation. The second disadvantage is that if Nyquist conditions are not satisfied, then the spectrum replicas will interfere with the spectrum of ideal image, see Fig. 5.7a. The Box filter will capture them and in reconstructed images aliasing effects will appear. Thus the solution is to use more soft low-pass filter, such as Gaussian filter which will attenuate the parasite replicas and the reconstruction image will be more smooth. The advantage of Gaussian is that both filter transfer functions in spectral domain and filter impulse response are Gaussians. The latter is expressed by the Eq. (5.6). Furthermore, the relation between scale parameters in pixel domain σ and Fourier domain σ_F is $\sigma \times \sigma_F = 1$, see e.g. in [Lin94]. Filtering by a Gaussian filter is illustrated in Fig. 5.7b.

5.3.2 Sub-sampling of Images and Features

Sub-sampling of images has been performed in the goal to reduce the tremendous quantity of pixels in the high-resolution input image and remove small details. Then the visual content understanding algorithms could be applied on a lower resolution images analyzing only significant, sufficiently large details in them. In visual content mining an example of such a philosophy as using lower-resolution representations have been used in the GIST descriptor proposed by Oliva and Torralba [OT01]. In the Rough Indexing paradigm we proposed [MBL04] a strongly sub-sampled version (eight times smaller than original) of video frames by a direct decoding from video compressed stream was used for detection of objects. Coming back to the sampling theory we briefly exposed in Sect. 5.3.1, the sampled digital image is ideally reconstructed image $I_R(x, y)$, Eq. (5.20) and its spectrum coincides with the spectrum of original image $I(x, y)$, see Fig. 5.7a. Now if we need to subsample the image further in order to reduce its size, then to respect the conditions of ideal reconstruction we need to reduce the bandwidth of its spectrum by sub-sampling factor in such a way that

$$\omega'_{xc} = \omega_{xc}/s, \, \omega'_{yc} = \omega_{yc}/s \qquad (5.23)$$

Here s is sub-sampling factor.

Such a reduction of bandwidth is realised by low-pass filtering of the image to sample and the most popular filtering method is to apply Gaussian filter for its "good properties" we explained in Sect. 5.3.1. A simple example is the dyadic sub-sampling, in which case a 3×3 Gaussian filter is applied to the original image and then each even raw and column are removed. An example of such a "Gaussian Pyramid" is presented in Fig. 5.8a. One can see that further we climb to the apex of the pyramid, less details are perceived in the visual scene but still it remains pleasant to observe, i.e. without aliasing effects and understandable.

In convolutional neural networks the convolution operation is performed with trainable filters some of them act as low-pass- and others as high-pass ones. In Fig. 5.8b we can see a kind of "high-pass" filter effect after the convolution (see second image from the bottom). This is why there is no sense to respect Nyquist conditions and sub-sampling is fulfilled quite arbitrary. It is called "pooling" operation (see Chap. 6 for more details), which is fulfilled by the maxpooling operator. The latter takes the maximum of feature values on predefined support, thus retaining the most significant features. Hence in the AlexNet architecture which feature layers are shown in Fig. 5.8b, already after the first convolution the feature maps are sub-sampled by the factor of 4. Then in subsequent layers convolution and pooling are not regular. We show some feature maps in Fig. 5.8 going to the apex of "feature pyramid". The final result of both chains of convolution and sub-sampling for the Gaussian pyramid and the AlexNet architecture is given in Fig. 5.8c. Obviously the results are different: in the left image despite its very low resolution we still can distinguish the structure of the original image, while the features displayed in the right image serve as input for a fully connected layer of the network for the final classification steps with regard to 1000 classes of ImageNet contest [KSH12] and convey quite another information.

5.4   Conclusion

Hence in this chapter we tried to trace an analogy between very well known operations in image processing such as convolution and sub-sampling and the propagation of a high-resolution image trough a Deep convolutional neural network. Despite the same convolution operation is used in both cases, in image processing we use a predefined filters while in a Deep CNN they are trainable with regard to the target classification task. The sub-sampling operation in image processing

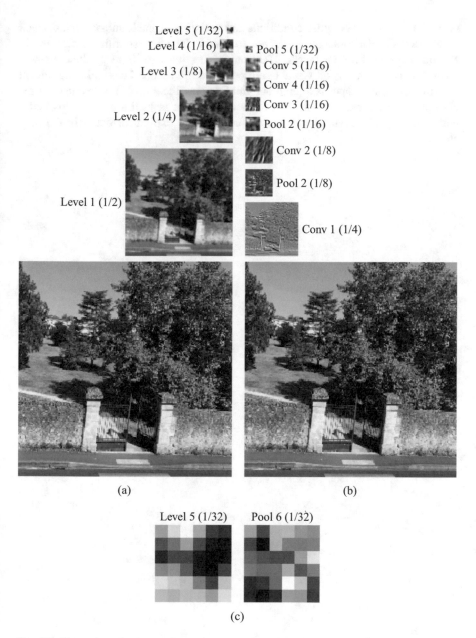

Fig. 5.8 Illustration of sub-sampling process: (**a**) by a Gaussian pyramid, (**b**) sub-sampling by convolution and pooling layers in AlexNet architecture [KSH12], (**c**) the output (6 × 6 pixels) of both pipelines

is well designed if Nyquist conditions are satisfied at each sub-sampling level. In a Deep CNN sub-sampling serves to retain the most significant features in feature maps of different layers. The goal in the first case is to produce a lower resolution image of a good visual quality, while in the second case all insignificant information from images is filtered out yielding classification of it into one of target classes. Nevertheless, we think that this comparison is justified and helps better understanding of convolutional neural networks we explain in detail in the next chapter.

Chapter 6
Convolutional Neural Networks as Image Analysis Tool

After studies of fundamental operations of convolution and sub-sampling in previous chapter, we introduce here convolutional neural networks and consider those designed for particular data: images. First of all we will expose some general principles, then go into detail layer-by-layer and finally briefly overview most popular convolutional neural networks architectures.

6.1 General Principles

Regular neural networks do not scale well for visual information mining. Their architecture, for a given neuron in a certain layer, connects all the activations from the previous layer to that neuron. In the context of Computer Vision, data takes the form of images (or videos), that is at least three dimensional objects. Let's imagine that we want to connect a neuron of the first layer to each pixel of a color image of rather small size $200 \times 200 \times 3$. This would give no less than 120,000 weights, just for one neuron in the first layer. Even shallow neural networks would become unmanageable regarding the number of weights (recall that each weight should be learned during the training).

On the other hand, it is crucial in image processing to consider the spatially-local correlation in images. For instance, if we wish to train an edge detector, it does not make sense to consider the whole image all at once for each neuron, we should rather decompose the image into different windows and apply the filter on each window, in order to make it react to windows with strong edges.

Convolutional neural networks (abbreviated CNNs) were specifically designed for Computer Vision tasks and their architecture presents differences compared to regular neural networks. The most obvious difference is that the layers of a CNN have their neurons arranged in three dimensions (height, width and depth) so as to match the geometric shape of the data (images can be seen as cuboids of pixels).

© The Author(s), under exclusive license to Springer Nature Switzerland AG 2020
A. Zemmari, J. Benois-Pineau, *Deep Learning in Mining of Visual Content*,
SpringerBriefs in Computer Science, https://doi.org/10.1007/978-3-030-34376-7_6

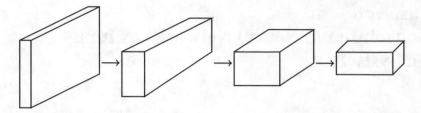

Fig. 6.1 In a CNN, the dimensions decrease layers after layers

In the Fig. 6.1, one can notice that the spatial dimensions decrease layers after layers, whereas the depth dimension increases. In most cases, the first layers of a CNN (sometimes called bottom layers) are locally-connected, that is their spatial support (also called receptive field) is limited which helps obtaining spatially-local correlation information. Furthermore, those first layers reduce the spatial dimensions by pooling operation. Once the spatial dimensions are small enough (when the top layers are reached), fully-connected layers are often used just like in a regular neural network, and the last layer is a fully-connected layer that computes the class scores.

The layers used in the bottom layers of CNNs are obviously characteristic of this particular type of neural network, due to their locally-connected property. Convolutional neural networks are indeed biologically-inspired: Hubel and Wiesel showed in [HW68] from the observation of animal's visual cortex that the latter consists of complex arrangements of cells, and that those cells are sensitive only to limited parts of the global visual field.

We give a brief description of the two most important locally-connected layers used in CNNs:

6.2   Convolutional Layers

Convolutional layers are at the core of the CNN architecture: they react to spatially-local correlation in input images. Those layers consist of a set of filters of limited spatial size, and those filters have their weights trainable. The name convolutional layer comes from the way this layer type behaves: it makes its filters slide across the whole image exactly as we compute convolution of an image with filter (see Chap. 5). On the contrary to the image processing operations the shift of filter mask when sliding along the input map can be greater than 1. This shift is called "stride" parameter and is a part of general settings or "hyper-parameters" of a Deep Convolutional Neural Network. At each position taken by a filter, a dot product is performed between the filter and the corresponding region in the image. Once all the filters are done sliding over the whole image, an activation map is obtained for each of those filters. Those activation maps are stored along the depth, which explains why the depth dimension tends to increase in CNNs layer after layer.

In practice, each filter learned by the CNN will activate when they encounter a specific visual feature. The filters tend to become more and more abstract when the layers go up. Filters from the bottom layers tend to react to simple objects such as edges, specific shapes or colors, whereas filters from the upper layers can react to more complex objects, such as buildings, animals, etc. depending on the dataset they were trained on.

In the sequel, we will give the intuition on how the use of CNN reduces the number of parameters (weights) of the network. For the sake of simplicity, consider that each image is encoded using a matrix whose elements are 0 or 1 as illustrated in Fig. 6.2.

Given a matrix (corresponding to an image), the convolution layers learn a set of filters (see Chap. 5). Figure 6.3 gives an example of the result of such operation when applied to the matrix of Fig. 6.2 with a filter of size 3×3 using a stride equal to 1.

Figure 6.3 explains how the convolution of an image (6×6) with a filter (3×3) and a stride equal to 1 is computed. We can observe that the resulting image is of smaller size (4×4). Filters correspond to the weights to learn and induce the connection between the input entries (of a layer) and the units of the convolution layer. This is illustrated in Fig. 6.4.

The key points are the following:

Fig. 6.2 Image encoding: (**a**) a handwritten digit and (**b**) its encoding

0	1	1	1	1	0
0	0	0	0	1	0
0	0	1	1	1	0
0	0	0	0	1	0
0	0	1	1	1	0
0	0	0	0	0	0

(a)   (b)

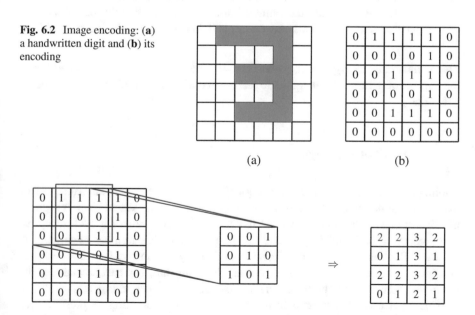

Fig. 6.3 The convolution of a matrix (6×6) corresponding to the image in Fig. 6.2 and a filter (3×3) and the resulting matrix

Fig. 6.4 The first two layers
of the CNN for the previous
example: the two units (red
and blue) are connected to a
subset of the input layer units
and not to all of them and the
two units share the same
weights for some of the
previous units

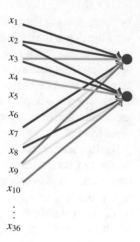

- the units of the convolution layer are not connected to all the units in the previous layer: when the filter is applied to the top-left corner of the matrix, it induces that the first unit (the red one) is connected to the corresponding entries: x_1, x_2, x_3, then x_7, x_8, x_9 and finally x_{14}, x_{15}, x_{16}.
- The units share the same weights: if the stride hyper-parameter is set to 1, then the filter is moved by one. Since we are still using the same filter, the weights for the connexions between the second unit (the blue one) and the corresponding entries (starting at x_2) are the same. Figure 6.4 explains this by the colors used (e.g. we have red connexions between x_1 and the red unit and between x_2 and the blue unit).

This is illustrated in Fig. 6.4. It is clear also that number of connections, and hence the number of parameters (weights) to learn is reduced compared to fully connected networks.

6.3   Max-Pooling Layers

Pooling reduces the computational complexity for the upper layers and summarizes the outputs of neighboring groups of neurons from the same kernel map. It reduces the size of each input feature map by the acquisition of a value for each receptive field of neurons of the next layer. Once again, this reduces the number of parameters to learn. Pooling is a general operation and many variants exist. However, we consider here the Max-Pooling operation, meaning that, for each rectangle, the maximum value is kept and the others are discarded as shown in Fig. 6.5.

This layer can be seen as a form of non-linear sub-sampling. It presents two interests:

- It reduces the spatial dimensions, which reduces the computational cost for the upper layers.

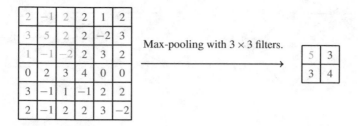

Max-pooling with 3×3 filters.

Fig. 6.5 An illustration of a max-pooling operation: the size of the image is reduced

- It provides a form of translation invariance. To see this, consider the case of a max-pooling on regions of size 2×2 followed by a convolutional layer. Regions can be translated by one pixel in eight directions. Among the eight configurations, three will produce the exact same output at the convolutional layer.

One should pay attention to the excessive use of pooling: indeed, reducing the spatial dimensions results in a loss of information which can affect the training. Note that there exists different types of pooling layers, such as average pooling and L^2-norm pooling, but the max pooling version has proven to be the most effective most of the time.

6.4 Dropout

One of the bottlenecks of supervised learning approach is the so-called "overfitting phenomenon". This means that the classifier after a training process classifies training data with a small error, but is not able to generalize well on unseen data. Overfitting can be avoided using model combination. This involves averaging the outputs of many separately trained neural networks which is extremely expensive, especially for deep neural networks.

Moreover, such type of averaging assumes (for effectiveness) that the averaged models are very different: they should either have different architectures or be trained on different data. The former is difficult to achieve because properly tuning an architecture is complex, and tuning many of them is even more. The latter is difficult to achieve for large models: deep neural networks indeed require more data to be properly trained, which means that different neural networks would probably exceed the available amount of training data.

The dropout method addresses those problems by sampling many thinned versions of the original neural network. Those versions are obtained by randomly discarding units and their connections from the neural network. More details can be found in [SHh+14]. The dropout regularization is illustrated in Fig. 6.6. We note that in CNNs Dropout is performed on Fully Connected layer and indeed increases network performances.

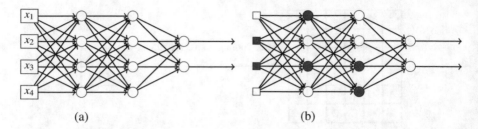

Fig. 6.6 Illustration of the dropout regularization: (**a**) the neural network and (**b**) the dropout application. At the beginning of any iteration in the training phase, each unit have probability p to be removed, together with its connections making the network to become thinner. During the test phase, if the dropout probability is p, each weight is multiplied by $1 - p$

6.5   Some Well-Known CNNs Architectures

It is extremely difficult to be exhaustive in the overview of popular architectures. We will limit ourselves to those which serve as a basis for quite a large set of applications.

6.5.1   *LeNet Architecture and MNIST Dataset*

LeNet is an extremely popular architecture, which was first introduced in 1998 by LeCun et al. [LBBH98]. This CNN architecture was designed to solve the problem of digit recognition in a very robust manner. It is a straightforward and rather shallow architecture (5 hidden layers only), making it a very popular CNN to understand the basics of Deep Learning.

The core principle of LeNet family architectures is the repetition of the pattern convolutional layer + activation + max-pooling layer in the lower layers. Resulting spatial reduction enables the use of cascading fully-connected layers for the upper layers, and a softmax classifier gives the score of the model.

It is very common to train the LeNet architecture over the MNIST dataset. The latter is a database of handwritten digits, containing a total of 70,000 digit images. This dataset splits the data into 60,000 images for training and 10,000 images for validation. Every digit image of this dataset has been size-normalized and centered in a 28×28 fixed-size image.

The simple architecture of LeNet combined with the compactness of the MNIST dataset are ideal to try new learning techniques and pattern-recognition methods on real-world data while spending minimal efforts on preprocessing and formatting. Excellent training results with accuracy exceeding 98% are possible to achieve in a very limited time, even without GPU acceleration.

The LeNet architecture is resilient to different types of transformations, thus it enables very robust character recognition. It provides (for reasonable transformations) many interesting properties, as illustrated in LeCun's website dedicated to LeNet [LeC]. Most notably, LeNet's robustness comes from the following properties:

- Translation invariance: this is mostly useful for vertical translations, as the positioning of the characters in a string is never perfect.
- Scale invariance: LeNet achieves this type of invariance over a wide range of sizes.
- Rotation invariance: the authors estimate that LeNet can recognize digits rotated by an angle of 40 degrees.
- Squeezing invariance: LeNet shows robustness to variations of the aspect ratio.
- Stroke width invariance: this type of robustness is useful to limit the need of unreliable preprocessing methods such as line thinning.
- Noise robustness: LeNet is resilient against various types of noise added above the digits.

6.5.2 AlexNet Architecture

Among the most popular architectures, AlexNet was presented in 2012 by Krizhevsky et al. [KSH12] in the paper entitled "ImageNet Classification with Deep Convolutional Networks", which is widely regarded as one of the most influential publications in the field of Deep Learning. This architecture is designed to solve a difficult classification problem on the dataset ImageNet.

The dataset ImageNet is an image database organized according to the WordNet hierarchy, which is a lexical database for English words. Currently only the nouns from WordNet are considered in ImageNet. Each node of the hierarchy is depicted by hundreds or even thousands of images, with an average of over five hundred images per node. This dataset is well known as it is extremely rich. An annual contest centered around ImageNet was created, the ILSVRC (ImageNet Large-Scale Visual Recognition Challenge), in which people can evaluate and compare their architectures by training their CNNs on the ImageNet database to solve object detection and image classification problems at large scale. The ILSVRC contest has seen many CNN candidates become staples in the field of Computer Vision over the past few years.

The authors presented the AlexNet architecture to the 2012 ILSVRC (ImageNet Large-Scale Visual Recognition Challenge) and won the contest by far with a top 5 test error rate of 15.4%, when the second best entry achieved an error of 26.2%. This result was seen as an outstanding performance and amazed the Deep Learning and Computer Vision communities.

The AlexNet architecture consists of 60 millions parameters for 500,000 neurons, 5 convolutional layers, some of them followed by a max-pooling layer and two fully-connected layers followed by a softmax classifier of size 1000 to score the model. We have illustrated this architecture in Fig. 5.3 in Chap. 5.

The authors also introduced regularization to their architecture under the form of dropout. As discussed in previous section, this method prevents overfitting even for very deep neural networks.

6.5.3   GoogLeNet

In 2015, Google published a new architecture [SLJ$^+$14]. GoogLeNet is a 22 layers CNN and was the winner of ILSVRC 2014 with a top 5 error rate of 6.7%. Along with its depth, what sets apart GoogleNet from most of the architectures back in 2014 is that it does not simply rely on an alternation of convolutional and pooling layers, which was the common usage since the introduction of the LeNet architecture. Instead of a sequential arrangement of the layers, the authors thought about the addition of parallelism in the architecture, with the introduction of 9 so-called "inception blocks" that are composed of different layers themselves, for a total of more than 100 layers. Figure 6.7 presents the architecture of an inception block.

GoogLeNet led the path to a new design philosophy in Deep Learning: the authors proved that a creative architecture could lead to improved performances and a computational efficiency. It opened the path to very creative CNN architecture designs.

6.5.4   Other Important Architectures

Many more CNNs architectures have proven to be staples in Deep Learning. Among them, we can cite:

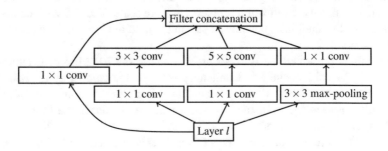

Fig. 6.7 GoogLeNet architecture: structure of an inception block

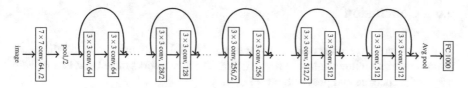

Fig. 6.8 Architecture of the ResNet network. It contains 34 layers and shortcuts each two convolutions

- ZF Net and DeConv Net (2013): Matthew Zeiler and Rob Fergus won the 2013 ILSVRC challenge with a CNN named ZF Net, which is a fine tuned version of AlexNet with ideas to improve overall performance. In their relative publication [ZF13], they also gave some rich insight regarding the intuition behind CNNs, and they also presented a special algorithm named DeConv Net which enables to visualize the response of the trained filters of a CNN when an image is fed to this CNN. This enables to examine what type of structures excite a given feature map and helps getting a deeper understanding of the way a specific architecture behaves.
- Microsoft ResNet (2015), see Fig. 6.8: Microsoft Research Asia proposed in 2015 a new architecture that broke many records [HZRS15]. ResNet was the deepest CNN when it was first presented, and it won the 2015 ILSVRC challenge, with an error rate of only 3.6%, which is usually lower than the error rate an average human would get. It introduced an architecture based on "residual blocks", which basically takes into account the input after a conv-relu-conv pattern, by adding the input to the output. According to the authors, "it is easier to optimize the residual mapping than to optimize the original, unreferenced mapping". Moreover, it helps dealing with the vanishing of the gradient. In many visual content recognition tasks ResNet with different amount of layers is used as the most efficient network. It was used in our example of research in Chap. 8.
- R-CNN (2013): In 2013, Ross Girshick and his group at UC Berkeley presented a new CNN architecture designed to solve the object recognition problem and called R-CNN [GDDM13]. The problem is split into two parts: the region proposal step and the classification step. For the first step, an algorithm of selective search is used [UvdSGS13]. This algorithm selects using bounding boxes a certain number of regions that have the highest probability of containing an object. These regions are then fed to a CNN, which outputs a feature vector for each region. Those vectors are then fed to a set of linear SVM algorithms trained for the classification problem. Note that this architecture was revised and accelerated in 2015 with the presentation of Fast R-CNN [Gir15] which optimizes the pipeline, and of Faster R-CNN [RHGS15] a bit later which simplifies the complex pipeline of R-CNN and Fast R-CNN, thanks to the introduction of a region proposal network.

6.6   Conclusion

In this chapter, we presented Convolutional Neural Networks, one of the most important architectures in image recognition. These networks have been extensively used in last decade and proved their efficiency.

However, when we have to deal with time and/or memory, CNN are not suitable and cannot be used for dynamic content mining. In next chapter, we introduce neural networks architectures that overcome these problems and thus can be used for videos, speech recognition, etc.

Chapter 7
Dynamic Content Mining

Neural networks and convolutional neural networks can be considered as functions which take as input a vector and compute a distribution over the set of possible classes. Such networks have no notion of order in time nor in memory. That is they are not suitable for dynamic content mining like speech recognition, video processing, etc.

Nowadays, we are "translating" into neural networks formalism all we know to do for visual content mining. For this task, Hidden Markov models were extensively used [KOG03, KBD$^+$14]. Hence, in this chapter, we discuss this formalism first and then we explain dynamic neural networks such as RNN and LSTM.

We thus present three approaches that can be used to deal with sequences and, hence, dynamic content: hidden Markov models, recurrent neural networks and long-short term memory networks.

7.1 Hidden Markov Models

Hidden Markov Models (HMM) are based on Markov chains. A *Markov chain* is a discrete-time stochastic process $(X_t)_{t \geq 0}$ s.t. each random variable X_t takes values in a discrete set S, called *state space*, and for any s, s' and $s_0, s_1, \cdots, s_{t-1} \in S$,

$$\Pr\left(X_{t+1} = s \mid X_t = s', X_{t-1} = s_{t-1}, \cdots, X_0 = s_0\right) = \Pr\left(X_{t+1} = s \mid X_t = s'\right). \quad (7.1)$$

If the set S is finite then the chain is said to be *finite-state*.

Equation (7.1) is called *memoryless* property and it simply means that, as time goes by, the process loses the memory of the past.

The chain is characterized by the space state S and by its *transition matrix* $P = \left(p_{i,j}\right)_{(s_i, s_j) \in S \times S}$, where,

© The Author(s), under exclusive license to Springer Nature Switzerland AG 2020
A. Zemmari, J. Benois-Pineau, *Deep Learning in Mining of Visual Content*,
SpringerBriefs in Computer Science, https://doi.org/10.1007/978-3-030-34376-7_7

$$p_{i,j} = \mathbb{Pr}\left(X_{t+1} = s_j \mid X_t = s_i\right), \forall t \geq 0, \text{ and } \forall (s_i, s_j) \in S \times S. \qquad (7.2)$$

Note that the transition matrix P verifies two properties: (1) its elements are all positive, and (2) each row sums to 1.

It is always possible to represent a finite-state Markov chain by a *transition graph* $G = (S, \tau)$ where S is the state space and τ corresponds to the transition matrix: for any pair of states s_i and s_j in S, $(s_i, s_j) \in \tau$ if and only if $p_{i,j} > 0$. The graph G is, thus, an oriented weighted graph. Given $t \geq 0$, the *distribution* at time t of the Markov chain is given by:

$$\pi_s^{(t)} = \mathbb{Pr}\left(X_t = s\right), \forall s \in S.$$

To characterize the chain completely, in addition to the state space S and the transition matrix P, one needs to specify the *initial distribution*:

$$\pi_s^{(0)} = \mathbb{Pr}\left(X_0 = s\right), \forall s \in S.$$

Thus, knowing $\pi^0 = \left(\pi_s^{(0)}\right)_{s \in S}$ and P, allows to compute $\pi^t = \left(\pi_s^{(t)}\right)_{s \in S}$. Indeed:

$$\pi^{(t)} = \pi^{(t-1)} P = \pi^{(0)} P^t, \forall t \geq 1.$$

A Markov chain is useful when we need to compute a probability for a sequence of *observable* events. In many cases, however, the events we are interested in are *hidden* and cannot be observed directly.

A hidden Markov model allows us to talk about both observed events and hidden events. Thus, to characterize a hidden Markov model, in addition to the state space S, the transition matrix P and the initial distribution π^0 as for Markov chains, we need to define two additional components:

- a set of T *observed* outputs $\{o_1, o_2, \cdots, o_T\}$, drawn from an output alphabet $V = \{v_1, v_2, \ldots, v_K\}$, i.e., $o_t \in V$, for any $t \in \{1, 2, \cdots, T\}$,
- A sequence B of *emission* probabilities: $b_i(o_j)$ corresponds to the probability of an observation o_j to be generated from state s_i.

Figure 7.1 presents a node in a graph corresponding to a hidden Markov model.

In addition to the memoryless property of Markov chains, we assume also the *output independence*: the probability of an output observation o_j depends only on the state s_i that produced the observation and not on any states or any other observations.

There are three fundamental questions we might ask of an HMM, [Rab89]:

- *Likelihood*: given an HMM $\mathbb{H} = (P, B)$, and an observation sequence O, compute the probability $\mathbb{Pr}\left(O \mid \mathbb{H}\right)$.
- *Decoding*: given an observation sequence x and an HMM $\mathbb{H} = (P, B)$, compute the best hidden state sequence s.

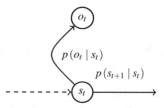

Fig. 7.1 A node in a hidden Markov model: possible transitions are from a hidden state $X_t = s_i$ to another one, and a node can emit an observation $o_t = o_j$. The conditional probabilities correspond to the transition and emission probabilities

- *Learning*: given an observation sequence x and the set of states in the HMM $\mathbb{H}$, learn the HMM parameters P and B.

7.1.1 Likelihood Computation

Given a Hidden Markov Model $\mathbb{H}$ and an observation O, the first problem is to compute the likelihood of O, i.e., $\Pr(O \mid \mathbb{H})$.

In the naive approach we compute the total probability of the observations O by summing over all possible hidden state sequences:

$$\Pr(O) = \sum_S \Pr(O \mid S) \times \Pr(S).$$

For an HMM with n hidden states and an observation sequence of T observations, there are n^T possible hidden sequences. When n and T are both large, n^T is very large, and we cannot compute the total observation likelihood by computing a separate observation likelihood for each hidden state sequence and then summing them. Instead of using such an exponential algorithm, we use an efficient $O(n^2 T)$ algorithm called the forward algorithm. The forward algorithm is a dynamic programming algorithm, and thus uses a table to store intermediate values as it builds up the probability of the observation sequence. The forward algorithm computes the observation probability by summing over the probabilities of all possible hidden state paths that could generate the observation sequence, but it does so efficiently by implicitly folding each of these paths into a single forward trellis.

Formally, if $O = o_1, o_2, \cdots, o_T$, then we introduce the forward variables recursively as follows:

$$\alpha_1(i) = \pi_i b_i(o_1), \tag{7.3}$$

and

$$\alpha_{t+1}(i) = \sum_{i=1}^{n} \alpha_t(i) p_{i,j} b_j(o_{t+1}). \tag{7.4}$$

Equation (7.3), *initialization*, means that the value of the first forward variable of state i is simply obtained by multiplying its initial probability by the (emission) probability of state i given the observation O at time 1.

The recursive equation (7.4), *recursion*, defines the forward variable of state j as the product of the previous forward variable of state i, multiplied by the transition probability $p_{i,j}$ between the previous state i to state j, multiplied by the emission probability from state j to the observation O.

Finally, the desired probability of an observation sequence O, given the HMM model $\mathbb{H}$, is obtained by summing up all the forward variables at time T, *termination*:

$$\Pr(O \mid \mathbb{H}) = \sum_{i=1}^{n} \alpha_T(i). \tag{7.5}$$

7.1.2 Decoding: The Viterbi Algorithm

Given an HMM $\mathbb{H}$, and a sequence of observations O, the decoding problem consists in finding the most probable sequence of hidden states s.

The Viterbi Algorithm [Vit67, Neu75] is a dynamic programming algorithm. It is similar to the forward algorithm used for the likelihood problem. It acts in four steps, initialization, recursion and termination, as in the forward algorithm plus an additional step named *backtracking*. The equations are also quite similar with slight differences.

The initialization step is exactly the same as Eq. (7.3), with different names for the variables:

$$v_1(i) = \pi_i b_i(o_1). \tag{7.6}$$

We also need to store the *backpointers* since the goal of decoding is to return the most likely state sequence. This is done by introducing the array variable bp:

$$bp_1(i) = 0. \tag{7.7}$$

In the recursion steps, Eq. (7.4) becomes:

$$v_{t+1}(i) = \max_{i=1}^{n} v_t(i) a_{i,j} b_j(o_{t+1}), \tag{7.8}$$

and one can observe that we are taking the maximum value, among the multiplication results, instead of the sum.

We also update, recursively, the backpointers:

$$bp_{t+1}(i) = \arg \max_{i=1}^{n} v_t(i) a_{i,j} b_j(o_{t+1}.) \tag{7.9}$$

The termination step is given by the equation:

$$p_* = \max_{i=1}^{n} v_T(i). \tag{7.10}$$

The value p_* represents the probability of the entire state sequence having been produced given the HMM and the observations.

We also compute the *start of the backtrace*:

$$s^* = \arg \max_{i=1}^{n} v_T(i). \tag{7.11}$$

The last step is the backtracking. It uses the backpointers arrays to find the hidden state sequence using the following equation:

$$s_t^* = bp_{t+1}\left(s_{t+1}^*\right). \tag{7.12}$$

7.1.3 Learning an HMM

The last question to ask of an HMM is: given a set of observations, what are the values of the initial distribution π, the state transition probabilities P and the output emission probabilities B that make the data most likely? We consider the number of states equal to n and the number of possible observations equal to K.

The standard algorithm to answer this question is the Baum-Welch algorithm [BPSW70] which is a special case of the Expectation-Maximization algorithm [DLR77]. It is an iterative algorithm which works as follows. We start with an initial probability estimates. This can be done using some prior knowledge about the parameters. Then we compute expectations of how often each transition/emission is used and then we re-estimate the probabilities based on those expectations. The process is repeated until convergence. More formally:

We first define the following variables:

$$\alpha_i(t) = \Pr(o_1, o_2, \cdots, o_t, s_t \mid \mathbb{H}), \tag{7.13}$$

and

$$\beta_i(t) = \Pr(o_{t+1}, o_{t+2}, \cdots, o_T \mid s_t, \mathbb{H}). \tag{7.14}$$

Then, one can define the probability (we skip the detailed calculations) of being in state i at time t and in state j at time $t + 1$, given the observed sequence O and the model $\mathbb{H}$:

$$\zeta_{i,j}(t) = \Pr(s_t = i, s_{t+1} = j \mid O, \mathbb{H})$$

$$= \frac{\alpha_i(t) p_{i,j} b_j(o_{t+1}) \beta_j(t+1)}{\sum_{i=1}^{n} \sum_{j=1}^{n} \alpha_i(t) p_{i,j} b_j(o_{t+1}) \beta_j(t+1)}. \tag{7.15}$$

Thus, we derive the probability of being at state i at time t, conditioned on the observation O and the model $\mathbb{H}$:

$$\gamma_i(t) = \Pr(s_t = i \mid O, \mathbb{H})$$

$$= \sum_{j=1}^{n} \zeta_{i,j}(t), \tag{7.16}$$

and, summing over observation O, we get:

- the expected number of transitions from state i:

$$\sum_{t=1}^{T-1} \gamma_i(t),$$

- the expected number of transitions from state i to state j:

$$\sum_{t=1}^{T-1} \zeta_{i,j}(t).$$

Then, we compute the current estimation of the desired parameters: (where $1_{x=y}$ is the indicator function, i.e., its value is 1 iff $x = y$)

$$\hat{p}_{i,j} = \frac{\sum_{t=1}^{T-1} \zeta_{i,j}(t)}{\sum_{t=1}^{T-1} \gamma_i(t)},$$

$$\hat{b}_i(o_j) = \frac{\sum_{t=1}^{T} 1_{o_t = o_j} \gamma_i(t)}{\sum_{t=1}^{T} \gamma_i(t)},$$

$$\hat{\pi}_i = \gamma_i(1). \tag{7.17}$$

The process is then repeated until it reaches the desired convergence.

7.2 Recurrent Neural Networks

As HMMs Recurrent neural networks (RNN for short) allow for sequence of prediction to be built when input sequence of observations is received. Recurrent neural networks (RNN for short) take as input the current example they see and what they have received previously. The decision of an RNN at time t is affected by the decision made in step time $t - 1$.

7.2.1 Definition

Recurrent neural networks have two sources of input: the present and the past. More formally, if we denote $y^{(t)}$ the response of the network at time t, then:

$$y^{(t)} = f\left(y^{(t-1)}, x^{(t)}\right).$$

Figure 7.2 explains intuitively the difference between RNNs and "classical" neural networks. In addition to the connections between nodes in a layer and the ones in its following layer, there are loops on the units of a layer.

From the discussion above, one can explain the idea behind recurrent neural networks: we aim to make use of sequential information. In the previous architectures, we assume that inputs and outputs are independent, i.e., the output at time t have no impact on the output at time $t + 1$. But for many machine learning tasks, and in particular for video mining, this is not true. If one wants to annotate a video frame, it is necessary to consider which frames have been seen before.

Recurrent neural networks were introduced to overcome this problem: they can perform tasks on sequences and thus the output at time t is considered in the input at time $t + 1$. We say that they have a memory which can be used to store information about what has been calculated so far.

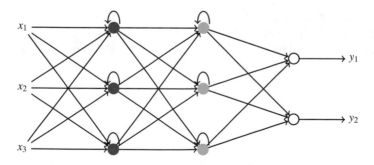

Fig. 7.2 A very simple representation of a recurrent neural network

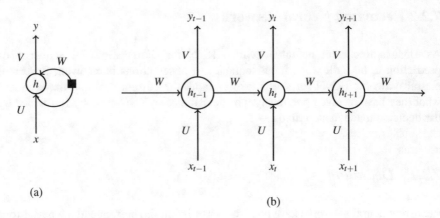

Fig. 7.3 A recurrent network and its unfolding

Figure 7.3a gives a simple representation of a neural network: the input x, combined with a matrix U (w.l.o.g., we consider the bias as a part of this matrix), is used to compute the value h of a hidden layer, then, this value is stored in the memory of the layer before it is passed to the output layer, using another matrix V.

Figure 7.3b gives another representation of the same concepts as in (a). It shows a recurrent neural network into a full network. This is called *unfolding*. By unfolding, we mean that we write out the network for the complete sequence. For example, if the sequence we care about is a video of 24 frames, the network would be unfolded into a 24-layer neural network, one layer for each frame. The following notation are used:

- x_t is the input at time t. This can be a one-hot vector encoding of a frame,
- h_t is the hidden state at time step t. It corresponds to the memory of the network. It is calculated using previous hidden value and the input at time step t:

$$h_t = f(Ux_t + Wh_{t-1}).$$

the function f is any activation function. It is usually a non-linear function such as tanh or *Relu*. The first hidden value h_0 is initialized to zeroes.

- y_t is the output at time step t. As for previous networks, it is expressed in terms of softmax:

$$y_t = \text{softmax}(Vh_t).$$

7.2.2 Training an RNN

As for any neural network training task, it is necessary to define a loss function which evaluates the performance of the network by comparing the predicted classes with the actual ones. For RNN, one can use any classical loss function (such as cross-entropy, etc.). Then, it is necessary to initialize the values of the parameters of the network (weights and biases). This step is dependent on the classification problem under consideration and on the input data. One can also use transfer learning to set the initial values. However, in practice, it is shown [SMDH13b, PMB13] that a Gaussian drawn initialisation with a standard deviation of 0.001 or 0.01 is a good choice.

Once the parameters initialized, it is necessary to use a method to train the network. Gradient descent based methods can be used. However, the time aspect in the structure of RNN makes classical methods (the ones we have seen for neural networks and for convolutional neural networks, see Chap. 4) not efficient. Thus, one needs to extend them to capture the time aspect of the RNN structure.

There are many methods that can be used to train an RNN. One can cite back-propagation through time [JH08], Kalman Filter-based learning methods [MJ98], . . . In this section, we will discuss the principle of the back-propagation through time (BPTT) method since it is an adaptation of the gradient descent method.

BPTT is a generalization of back-propagation for feed-forward networks. The main idea of the standard BPTT method for recurrent neural networks is to unfold the network in time (see Fig. 7.3) and propagate error signals backwards through time. As explained in Fig. 7.3, the set of parameters θ is defined by W, U, V and of course the set of biases (which, in our case, are encoded in the matrices). Computing the derivatives of the loss function with respect to V is roughly the same as for feed-forward networks. The extension is mainly done for the computations of the derivatives for U and W. Indeed, it is necessary to sum up the contributions of each time step to the gradient. In other words, because W is used in every step up to the output we care about, we need to backpropagate gradients through the network all the way.

Recurrent neural networks have difficulties learning long-time dependencies. This is due to the vanishing of the gradient. This happens when the derivatives of the activation function (sigmoid or tanh) approach 0. The corresponding neuron is then saturated and drive other gradients in previous layers towards 0. The small values in the matrices and the multiple matrix multiplications yield to a very fast gradient vanishing. Intuitively, this means that the gradient contributions from far away steps become 0. This problem is not limited to recurrent neural networks and is also present in (very) deep feed-forward networks.

There are solutions to avoid vanishing gradient problem. The most common solution is to use the *ReLu* as an activation function instead of the sigmoid or the tanh functions. A more suitable solution is to use the Long-Short Term Memory (LSTM) networks.

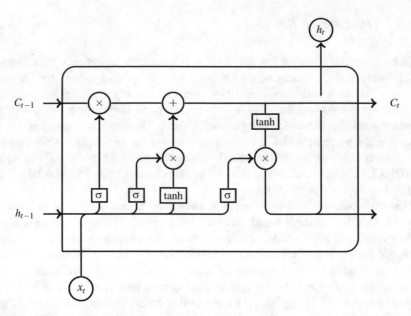

Fig. 7.4 A unit of a Short Long-Term Memory network: σ and tanh are respectively sigmoid and tanh functions as introduced in Chap. 3. Operators $\times$ and $+$ are simply the multiplication and addition operations

7.3 Long-Short Term Memory Networks

7.3.1 The Model

Recurrent networks are useful to capture the sequential dependencies in the inputs. However, as explained in the previous section, training such networks suffers from the exploding or the vanishing of the gradient. Thus the recurrent networks are not sufficiently efficient when used to learn long-term sequential dependencies in data. In 1997, Hochreiter and Schmidhuber [HS97] introduce the *Long Short-Term Memory* networks (LSTM for short). This architecture, and its variants become largely used in last years and proved its efficiency in many fields including speech recognition, text processing etc.

LSTM is a recurrent neural network where the units are more complex than in simple RNN. Figure 7.4 presents a typical unit of an LSTM.

Information comes to the unit from the current data x_t and from the previous value h_{t-1} of the hidden layer. The unit computes the value of the *forget gate*:

$$f_t = \sigma\left(W_f.[h_{t-1}, x_t] + b_f\right).$$

This corresponds to the information which is going to be thrown from the cell state C_t.

The same computation is done for the *input gate*:

$$i_t = \sigma\left(W_i.[h_{t-1}, x_t] + b_i\right),$$

and another value:

$$c_t = \tanh\left(W_C.[h_{t-1}, x_t] + b_C\right).$$

These values are then used to compute the new value C_t of the cell *state*:

$$C_t = f_t C_{t-1} + i_t c_t.$$

The last thing to decide is h_t, the output of the cell. This is done in two steps. First we compute the value of the (intermediate) output:

$$o_t = \sigma\left(W_o.[h_{t-1}, x_t] + b_o\right),$$

then:

$$h_t = o_t \tanh(C_t).$$

There are many other variants of the LSTM networks. The *Gated Recurrent Unit*, or GRU, introduced by Cho, et al. [CVMBB14] aim to simplify the computation and combines the forget and the input gates into a single gate: the *update gate*. It also merges the cell state and the hidden state.

7.4 Conclusion

In this chapter, we presented techniques used for dynamic visual content mining. We first introduced Hidden Markov Models, we discussed the main questions that this models can solve. Then, we discussed Recurrent Neural Networks and Long Short-Term Memory architectures, very powerful tools used for sequences and dynamic contents. We explained how these models can be trained by unfolding in time their underlying networks.

Chapter 8
Case Study for Digital Cultural Content Mining

In this chapter we consider an application case of Deep Learning in the task of architectural recognition. The main objective is to identify both: architectural styles and specific architectural structures. We are interested in attention mechanisms in Deep CNNs and explain how real visual attention maps built upon human gaze fixations can help in the training of deep neural networks.

8.1 Introduction

The overgrowing performances of Deep NNs allow for an application of them in real-life visual content mining tasks among which organization and interpretation of digital collections of cultural heritage is probably one of the most fascinating. This is why our case study is devoted to recognition of rich cultural content assets. We describe our research on classification of buildings of different styles in Mexican cultural heritage. But this is not the only point of our focus. When interpreting content of interest, humans are attracted by both natural contrasts and colours, and specific details of the target scenes. Our visual attention is selective: to recognize visual scene we foveate at the areas of interest. Hence when working with such an interesting visual content it is natural to try to incorporate visual attention mechanisms in powerful classifiers: Deep NNs. Specifically during the process of writing this book, the interest of community shifted to the so-called "attentional mechanisms" in Deep NNs [JLLT18], for text mining, translation or image classification. Hence, we are also interested in such mechanisms in the NNs in here, but try to relate them more to the human visual interpretation of scenes of interest. In the follow up, we will first review some methods relevant to our task of architectural objects recognition and then will try to see what attention mechanisms are the most efficient: those induced by the NNs themselves or those explicitly induced by human visual experience.

© The Author(s), under exclusive license to Springer Nature Switzerland AG 2020
A. Zemmari, J. Benois-Pineau, *Deep Learning in Mining of Visual Content*,
SpringerBriefs in Computer Science, https://doi.org/10.1007/978-3-030-34376-7_8

The recognition of architectures or the recognition of architectural styles has been addressed using different techniques from the areas of matching, pattern recognition and machine learning. Ali et al. [ASJ+07], present one of the first approaches for the identification of architectural structures, introducing a classifier system that provides the detection and location of windows as the basis for a post processing step. The goal is to bring this development to several systems, as mobile computing and provide areas of interest for a more complex post processing. The identification is based on the method of Viola and Jones [VJ01], inspired by the problem of face detection but trainable to detect a variety of objects in rates of real-time applications. This approach is a simple and efficient technique to perform a content description to propose regions that can define a structure through the spatial distribution of building's windows.

From a different perspective, Berg et al. [BGM07] parse images containing architectural structures. The output is a pixel-labeled image for the identification of the roof, vegetation, windows, boundaries of the buildings and doors, using a probabilistic model trained with color and texture descriptors. The database is confirmed by 200 manually pixel-annotated images and most of them refer to architectural elements.

Later, Zhang et al. [ZSGZ10] established that breaking up the image into regions and processing each region in a multiresolution approach can be better than using the whole image for recognition of architectural elements of Chinese culture. Each region is described by Histogram of Gradients (HOG) features at multiple resolutions (multi-scale) with Principal Component Analysis (PCA) dimensionality reduction to train five classifiers: (i) Support Vector Machine (SVM), (ii) K-Nearest Neighbor (K-NN), (iii) Nearest Mean Classifier (NMC), (iv) Fisher Linear Discriminant (FLD) and (v) a Multi-layer Perceptron (MLP). Cross-validation test indicates that the SVM performance was over 60% compared to MLP accuracy of 30% and K-NN's accuracy of 50%.

Mathias et al. [MMW+11] propose a pipeline to process the facades and determine the architectural style; Haussmann, Neoclassical and Renaissance by using the Scale-Invariant Feature Transform (SIFT) descriptor. The method contains several stages: (i) it starts with a scene classification (no buildings, building part, street and facades), (ii) the orientation of the facade is rectified (orthogonal to the camera), (iii) the buildings are split into single ones (there might be different styles in a single block), (iv) characteristics are extracted and finally, (v) a Naive-Bayes classifier is trained. The classification rate is above 85%, for the identification of facades a part of a building and the street. Rejection class accuracy or "no-building" is 100% for proposed dataset.

Shalunts et al. [SHS11] argue that to identify architectural structures it is indispensable to identify several elements separately. The identification of the windows corresponds to three classes; Romanesque, Gothic and Baroque. They use Bag of Visual Words (BoW) through image histograms as classifier. The database consists of only 400 images taken by the authors and retrieved from Flickr platform [Fli04]. The accuracy of this model is around 95.33%, based on the presented confusion matrix for three-class identification.

Next, Shalunts et al. [SHS12], also proposed the identification of Gothic and Baroque architectural elements. The proposed method seeks the identification of unique elements from each architectural structure as in [SHS11], such as windows and ornaments from the facade. The database is conformed of 520 images in two classes (Gothic and Baroque). They propose the use of SIFT descriptor and BoW as model.

Xu et al. [XTZ+14] report an algorithm Multinomial Latent Logistic Regression (MLR) as a deformable part model based on SVMs and applied to a specific-purpose database consisting of 25 architectural structures. The method is based on the use of HOG characteristics at different scales, defined as a pyramid. The model includes latent variables with the objective of obtaining by regression the class of a given image.

Shalunts [Sha15] shows the identification of three styles of towers, with features Romanesque, Gothic and Baroque. As a descriptor, the authors use SIFT and K-NN for the classification stage. The effective recognition of the proposal is 80.54% of accuracy. A specific-purpose database is presented, created by the author and Flickr images.

In essence, in the last years several methods have been proposed where classifiers and feature-based methods are used for the identification of structural elements. Deep learning approaches have showed their increased performance compared to feature-based methods. Therefore, they are interesting to be applied for such a recognition task. Furthermore, following the recent trend in the Deep CNNs such as attention mechanisms to re-inforce some features in the maps through the network, we wish to incorporate saliency of pixels and features in the network. Hence we will first focus on visual saliency and explain how we understand it in our task.

8.2 Visual Saliency

Visual saliency is a measure of attraction of humans by specific details in the visual scenes. There are fundamentally two kinds of visual attention and thus the derived saliency: bottom-up and top-down. Bottom-up, or stimuli driven expresses the attraction of human visual system by contrasts, colours, changes in orientation, residual motion in video. The second one—top down, is task driven and depends on the target visual task. More detailed explanations and examples of the use of bottom-up saliency in visual indexing tasks can be found in previous book [BPC17]. In this work we will focus in top-down saliency, as visual task is particular.

8.2.1 Top-Down Saliency Maps Built Upon Gaze Fixations

The generation of top-down saliency maps is supported by a psycho-visual experiment. The goal of this psycho-visual experiment consists in recording gaze fixations of subjects in the visual task of recognition of architectural styles of displayed buildings. It was designed accordingly to the methodology for task-driven goal-oriented visual experiment [Duc07]. We have selected a group of subjects of similar educational level (Master and PhD students) of 23.7 ± 2.8 years old.

Gaze fixations provide information to build a ground truth reference about visual attention for specific object recognition task. In this case, the ground truth is a subjective saliency map built from eye-tracking measurements in a subset of images of Mexculture Architecture Dataset [ORR$^+$16].

Using an eye-tracker system, we record where participants are looking during the specific task of building style recognition. Eye-tracking data are recorded at a regular rate (250 Hz in our case) in the coordinates of the eye-tracker system, where the origin is the center of the experimental screen. Then, measurements are transposed to frame coordinates where the origin is the top left corner to finally compute subjective saliency maps. Gaze fixations from a single participant are not representative to describe the visual attention. Then, the subjective saliency maps are built with gaze fixations of multiple participants. The resulting map should provide information about fixations density.

David S. Wooding [Woo02] proposed a method to generate the saliency maps in three steps: (i) for each participant a partial saliency map $S_{subj}(I)$ is build applying a Gaussian (depicting the fovea region on the screen) at each measurement position, (ii) all partial subjective maps are summed up to form $S_{subj}'(I)$ and (iii) the final saliency maps is normalized and stored as $S_{subj}(I)$.

More formally, to compute single Gaussians on gaze-fixations a Gaussian spread σ is computed from a fixed angle $\alpha = 2°$ to estimate σ_{mm} based on the eye-screen distance $D = 3H$, where H is the screen height.

$$\sigma_{mm} = D \times tan(\alpha) \tag{8.1}$$

Measures in mm are converted to pixels based on the screen resolution, where R is the screen resolution in pixels per mm,

$$\sigma = R \times \sigma_{mm}. \tag{8.2}$$

First, for each gaze fixation measurement m of the image I a partial saliency map is computed,

$$S_{subj}(I, m) = A e^{-\left(\frac{(x - x_{0m})^2}{2\sigma_x^2} + \frac{(y - y_{0m})^2}{2\sigma_y^2}\right)}, \tag{8.3}$$

with $\sigma_x = \sigma_y = \sigma$ and $A = 1$.

At the second step, all the partial saliency maps are summed up in $S_{subj}'(I)$,

$$S_{subj}'(I) = \sum_{m=0}^{N_m} S_{subj}(I, m) \tag{8.4}$$

Fig. 8.1 Gaze fixations and subjective saliency map generated: (**a**) Gaze fixations over the image and (**b**) saliency map over the source image. Red: more attention, Blue: less attention

(a)   (b)

where N_m is the number of gaze fixations recorded on all participants for the input image I.

Finally, the saliency map is normalized by the $argmax$ of $S_{subj}'(I)$ and stored in $S_{subj}(I)$,

$$S_{subj}(I) = \frac{S_{subj}'(I)}{argmax(S_{subj}'(I))}. \tag{8.5}$$

An example of gaze fixations and the generated subjective saliency map are shown in Fig. 8.1.

8.2.2 Co-saliency Propagation

To use specific saliency maps in large database is not always possible as this requires a large scale psycho-visual experiment, when subjects have to observe images along the day. This is why in [MOBG$^+$18] a co-saliency propagation method was proposed. This method is adapted to the databases where several instances of the same objects are present. This is the case of architectural databases, or crowd-sourced image collections when different subjects input the images of the same object, but shoot in different conditions and from different points of view. The method requires a recording of gaze fixations not on the whole database, but only on a "boot-strap" subset. Given a subset of reference images and their subjective saliency maps, the co-saliency propagation is based on the assumption that: if a subject is looking at particular details of an image for specific object identification, then he/she will most probably seek for the same details in other representations of the same object in different images, even if the object is seen

Fig. 8.2 Reference and projected saliency map of reference and target images by matching points. A red line depicts keypoints matching and green polygon over target image shows the transformation given by the estimated homography. Here, the parameters are Lowe's ratio and $\alpha = 20°$ as the angle threshold for the homography validation step. Red: more attention, Blue: less attention

from a different point of view. Hence the idea here consists in estimating the homography H between two images, the reference I_r, with available gaze fixations and the target I, which was not shown to subjects during the psycho-visual experiment. Then with the help of estimated homography H, the coordinates of the gaze-fixations in I_r are translated into the plane of the image I. After that the "co-saliency" map is computed accordingly to the equations presented in Sect. 8.2.1. To compute the homography a SIFT keypoints matching is performed and then the homography H is computed with the RANSAC algorithm on matching keypoints [FB81].

The co-saliency propagation method is illustrated in Fig. 8.2. To propagate subjective saliency maps to similar images, we find matching keypoints in both, reference and target images. Using SIFT keypoints, we estimate the homography from reference image to target image and transform gaze data to target perspective, allowing computation of co-saliency maps for target images. When matching keypoints over real data some issues appear due to different context changes, object scaling and contrast changes between samples. Due to this, the homography is not always correct, the more matching points we have, the more accurate the homography estimation is. Then, to discard wrong perspective predictions we perform a geometry test of the resulting projection of reference image on target image plane. Transforming the corners of the reference image with the estimated homography, we should always obtain a convex four sided polygon, which means by definition that each internal angle is under 180°. Obviously, small variations of the transformed reference coordinates are present because the homography plane transformation. We set up an internal angle hysteresis ($\alpha = 20°$) allowing only a limited perspective changes, discarding very wrong projections.

8.3 Saliency in Deep CNNs

8.3.1 Attentional Mechanisms in Deep NNs

In this section we present Deep NNs with attention from literature *Squeeze-and-Excitation Networks* [FSM+17], *Double Attention Networks* and the network with forward-backward external visual attention propagation we propose.

All the attention modules we integrate in our experiments can be constructed given the input $\mathbf{X} \in \mathbb{R}^{H \times W \times C}$, with exception that *visual saliency-based attention* block requires also $S \in \mathbb{R}^{H \times W \times 1}$, the saliency map for the current image sample. Here, $\mathbf{X}$ denotes a set of features, product of a previous layer, normally a convolutional layer. These different mechanisms are presented in the following subsection.

8.3.2 Squeeze-and-Excitation Networks

This kind of networks were first introduced with the aim of improving channel inter-dependencies like a channel-attention block, maintaining a very low computational cost [FSM+17]. The core idea is to add parameters to each channel after convolutional blocks, thus, the network weights the relevance of each feature map during optimization process. The most basic form of this method, can be seen by assigning a single trainable parameter associated to each channel in the input features map, giving a strong reference on how relevant the channel is. This method can be exemplified in two steps, the first one is to obtain a global perception of each channel, squeezing feature channel maps in a single value by global pooling, the result is a vector of n-dimension, where n is the number of channels in the input feature map. As shown in Fig. 8.3, each channel is summarized by $\mathbf{F}_{\text{squeeze}}(\cdot)$. Then, the second step is to feed a two-layer fully connected neural network which produces a vector of the same size, denoted by $\mathbf{F}_{\text{excitation}}(\cdot, \mathbf{W})$ in Fig. 8.3. Finally, these n values are used to weight each channel in the set of original features.

The Squeeze-and-Excitation block is a two-layer depth fully connected neural network and it requires two parameters to be set: the number of C the input and the ratio r. Those parameters are used to define the number of weights the first layer must contain, where its output is shaped as $1 \times 1 \times \frac{C}{r}$ and the output of the second layer is shaped as $1 \times 1 \times C$, where C is the number of channels.

8.3.3 Double Attention Networks (A^2-Nets)

The double attention mechanism is designed to capture long-range feature inter-dependencies by gathering and distributing features [CKL+18]. The idea, motivated

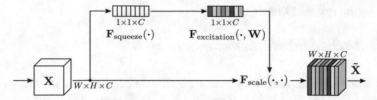

Fig. 8.3 Squeeze-and-Excitation block, re-drawn from [FSM+17]

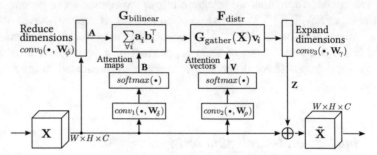

Fig. 8.4 A^2: double attention block, re-drawn from [CKL+18]

by *squeeze-and-excitation networks*, is to first collect features from the entire space and then distribute them into each location of the input, conditioned on a local feature vector for the given position. The network produces the "attention maps" and "attention vectors". The latter are obtained from current input feature map by convolution of it, see $\mathbf{W}_\delta$ and $\mathbf{W}_\rho$ filters in Fig. 8.4. Then the softmax operator enforces strong features and normalizes all of them between 0 and 1. This operation gives a sort of "importance" or "attention" map which is convolved with input features in Gathering block and multiplied with gathered features in Distribution block. Hence, the first attention mechanism is able to gather features from the source and the second attention mechanism is in charge of adaptively distributing information given the local attention vector. The output of both mechanisms aggregates information to each location in the input feature maps.

On one hand, as shown in Fig. 8.4, the task of gathering features is performed by computing a sum of pooling of second-order features given the outer-product of a pair of feature maps. This is known as bilinear pooling, in [LRM15] it is used to capture second-order statistics from data and generate better global representations. Features gathering is performed as,

$$\mathbf{G}_{\text{bilinear}} = \mathbf{AB}^\top = \sum_{\forall i} \mathbf{a}_i \mathbf{b}_i^\top \tag{8.6}$$

where $\mathbf{a}_i$ and $\mathbf{b}_i$ are *hwc*-shaped vectors (features) coming from the convolutional layers $\phi(\mathbf{X}, \mathbf{W}_\phi)$ and $\delta(\mathbf{X}, \mathbf{W}_\delta)$, respectively.

The authors claim that this first step is an effective manner to capture key global-features, such as texture and lighting. On the other hand, the second attention step is to distribute features back to different locations of the input in given locations $\mathbf{v}_i$ as,

$$\mathbf{F}_{\text{distr}} = \mathbf{G}_{\text{gather}}(\mathbf{X})\mathbf{v}_i \tag{8.7}$$

where $\mathbf{v}_i$ is given by $\mathbf{V} = \text{softmax}(\rho(\mathbf{X}, \mathbf{W}_\rho))$. The formulation of this second mechanism as a soft attention is ensured by selecting a subset of feature vectors from the first step with soft attention.

Finally, the double attention block combines the two above mentioned attention mechanisms into a single one with the aim of adding information to the original input as,

$$\tilde{X} = \mathbf{X} + \mathbf{F}_{\text{distr}}(\mathbf{G}_{\text{gather}}(\mathbf{X}), \mathbf{V}). \tag{8.8}$$

Unlike squeeze and excitation blocks, this block of double attention focuses on spatially identifying the interdependencies that may exist in the feature maps and not only measures the relevancy of each channel, but also adds spatial information relevant to the scene.

8.3.4 Visual Saliency Propagation

Hence after analyzing the so-called "attention models" in Deep NNs we will now show how we can introduce visual saliency from human observations of visual content into the whole CNN architecture whatever it be for scene interpretation tasks. The available gaze fixation maps will

- guide selection of features in pooling layers,
- control drop-out process for regularization of network,
- influence back-propagation in parameter optimization.

We thus introduce visual saliency in all blocks of a network architecture dealing with features and parameters. We are speaking about *end-to-end visual saliency propagation* through a Deep Convolutional Neural Network.

8.3.4.1 Saliency in Pooling Layers

A pooling function outputs a summary of its input based on specific conditions. Each element in the output comes from a neighborhood around the value of interest. Formally, the pooling layer takes the feature map $\mathbf{F}$ as input and produces the feature map $\mathbf{F}_m$ of reduced resolution, which depends on stride parameter (see Chap. 6). The receptive field in pooling layers, as in other visual layers is given by the kernel size.

Max Pooling   The max pooling, outputs the maximum value of each neighborhood to represent each values from the input image,

$$F_m(x, y) = \max\{F(\bar{x}, \bar{y})\}, \tag{8.9}$$

where $(\bar{x}, \bar{y})$ denotes the neighborhood around (x, y). Points (x, y) are regularly sampled in input features map with parameterized stride value.

Max pooling is the most used method of sub-sampling in CNNs, it produces a compact representation of input features, giving robustness to small image translations and reduction of dimensionality, which represents less parameters to train in deeper layers. But what about saliency-based pooling? We have introduced it in [MOBPGVRA18].

Saliency-Based Pooling of Features   As a method for selection of relevant regions in pooling layers, we use a pruning process of visual data for high saliency areas pooling. Our method takes origin from González-Díaz et al. [GDBBP16] and Vig et al. [VDC12]. We pool regions based on a random sampling process following a cumulative distribution function (CDF) of the Weibull distribution and the saliency map associated to the input image.

As in [VDC12], we consider saliency value of pixels to follow a cumulative Weibull distribution,

$$P(r; k, \lambda) = 1 - e^{(r/\lambda)^k}, \tag{8.10}$$

and a random variable S associated with saliency, the probability $P(S \leq r)$ is given by Eq. (8.10), where $r \in [0, 1]$ is an uniformly distributed random variable, $k > 0$ gives the distribution shape and $\lambda > 0$ the scale factor of the Weibull distribution.

Taking as input the features map $\mathbf{F}$ and a saliency map $\mathbf{S}_m$ and following the rule that regions will be pruned in function of $P(S \leq r)$, the random pooling strategy is defined as,

$$F_m(x, y) = \begin{cases} \max\{F(\bar{x}, \bar{y})\} & \text{if } S_m(x, y) \geq P(S \leq r) \\ \min\{F(\bar{x}, \bar{y})\} & \text{otherwise,} \end{cases} \tag{8.11}$$

where as in Eq. (8.9), $(\bar{x}, \bar{y})$ denotes the neighborhood around regularly sampled pixel value in (x, y). Hence, in highly salient regions of the input features map the baseline max-pooling is applied, in other regions minimal pooling will suppress the features.

As shown in Fig. 8.5, k controls the influence of the random sampling for region pooling. For lower values of k the CDF takes a shape where low saliency values have bigger probabilities to survive the pruning process. The shape of the CDF for bigger values of k gives a balance between low and high values of saliency for regions pooling. Besides, λ limits the minimum area to be processed, by shifting the CDF within the range [0,1]. Analyzing Fig. 8.5 again, we state that for a low value of k

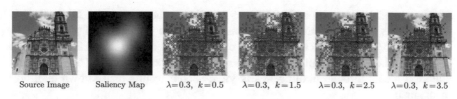

| Source Image | Saliency Map | $\lambda=0.3,\ k=0.5$ | $\lambda=0.3,\ k=1.5$ | $\lambda=0.3,\ k=2.5$ | $\lambda=0.3,\ k=3.5$ |

Fig. 8.5 Random pooling of features for different values of k-parameter; original image is given as a feature map with its respective saliency map

Fig. 8.6 Saliency maps introduced in pooling layers. Feature maps are not resized for a better visualization

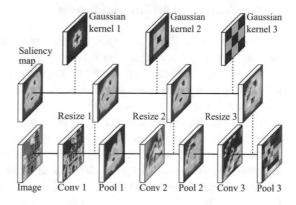

the concentration of pooling will be sparse, taking regions out of the highest values of the visual attention map. Thus, for a better selection of features to pool, we fixed $\lambda = 0.3$ and $k = 3.5$.

Saliency maps thus have to be introduced in all pooling layers of the network, as illustrated in Fig. 8.6. And for each layer they have to correspond to layer dimensions. Hence, taking a predicted saliency map for the input layer of the network, we resize it by conventional Gaussian low-pass filtering and sub-sampling. The scale parameter σ of Gaussian filter in each layer is computed accordingly to the filter size $l = \rho - 1$ with ρ the "stride" hyperparameter of the current layer.

As shown in Fig. 8.6, we use the image for training and the saliency map acts as a support in pooling layers.

Accordingly to the experiments fulfilled in [MOBPGVRA18] introduction of saliency-based feature selection in pooling layers, yield better stability in training of networks and slightly increases accuracies. Hence for AlexNet [KSH12] on the Mexculture Architecture Dataset [ORR$^+$16], the optimal model was obtained at 16,680th iteration vs 20,433th iteration for the base line model when using batch size of 230 images. The accuracy increase with the optimal model thus trained on validation dataset was 94.48% with regard to 94.45% for the baseline. We note that in this work, an automatic visual saliency prediction model that one of Harrel was used [HKP07]. The optimization method used was stochastic gradient descent with Nesterov's momentum. If we compare the performance of visual saliency-based pooling in CNNs vs "attentional mechanisms", such as squeeze and excitation networks and double attention networks we have presented in Sects. 8.3.2 and 8.3.3, then we stated the accuracy of 96.35% vs 92.36% for squeeze and 89.57% for double attention networks, respectively. All these models were implemented with ResNet-26 [HZRS15] which gave an accuracy of 92.19% as a base-line. Hence we can claim, that the NN does not learn everything and the propagation of a valid visual attention model through pooling layers yields better results than known attention mechanisms.

8.3.4.2   Saliency in the Dropout Layers

Drop out layers in CNNs serve for regularization, as it was described in Chap. 6. Regular dropout is performed on fully-connected (FC) layers (see Chap. 6). The question is how we can drop synaptic connections not completely randomly, but privileging those connecting salient features in the receptive field with the input of a neuron? Hence, here if we wish to use saliency maps propagated through the network, then we cannot use the FC layers. Indeed, each input neuron in an FC layer takes *all* neuronal responses from the previous layer. It is difficult to "kill" non salient neuron from one FC layer to another, as the notion of saliency is lost, hence we proposed the so-called spatial dropout of synaptic connections from neurons of the last convolutional layer and first FC layer. Indeed, spatial arrangements or features is preserved here and corresponding saliency map can be used.

Inspired by our previous work [MOBPGVRA18], instead of using a uniform or flat Gaussian distribution for dropping synaptic connections, [GBCB16] we use the visual attention map which is propagated through the network and is considered following Weibull distribution function.

The adaptive spatial saliency dropping starts with the cumulative Weibull distribution function (CWDF) defined by Eq. (8.10).

The goal is to provide a similar behavior of spatial saliency dropping at each processed feature map. Based on Eq. (8.10), first, we fix k based on experimental tests. Then, we find λ^* to ensure that the maximum value of the CWDF will be reached when $x = \mu_{sm}$. Where μ_{sm}, is the mean value of the saliency map of a given input image. The value of λ^* must satisfy,

$$e^{-(\mu_{sm}/\lambda^*)^k} = 0.0001, \qquad (8.12)$$

then,

$$\lambda^* = \frac{\mu_{sm}}{\log{(0.0001)^{\frac{1}{k}}}}, \qquad (8.13)$$

thus, $P(r; k, \lambda^*)$ will be very close to 1 when $r = \mu_{sm}$. We set 0.0001 in Eq. (8.12) to approximate the exponential part of Eq. (8.10) to zero.

Once we compute λ^* we have the shape of the CWDF to randomize spatial dropping only where values of visual attention map are under μ_{sm}, giving priority to activations to survive if they are inside the attention region.

Following the CWDF in Eq. (8.10), previously adapted with Eq. (8.13) and k, we define a new random variable s which follows a uniform distribution. We drop activations mapping s, through inverted adapted CWDF, and comparing the response against values of the saliency map S_m. Then, the response is defined as follows,

$$R(s) = \lambda^* \sqrt[k]{-log(1-s)} \qquad (8.14)$$

Activations map A_m is spatially dropped then as,

$$A_m(x, y) = \begin{cases} A_m(x, y), & \text{if } R(s) < S_m(x, y), \\ 0, & \text{otherwise,} \end{cases} \qquad (8.15)$$

where $R(s)$ is the response of inverse CWDF for dropping activations in $A_m(x, y)$. Random variable s is different for each $A_m(x, y)$.

The influence of k in Eq. (8.10), gives the possibility of adjusting the behavior of dropping activations where the values of the saliency map are under μ_{sm}. If the value of k is high, less activations are dropped when they are close to μ_{sm}. This behavior is shown in Fig. 8.7 with k chosen accordingly, where we take the saliency map as a feature map to visualize better how activations are dropped in regions close to μ_{sm}, λ parameter is automatically computed from the saliency map. Thus introduced visual saliency is used in a "hard way" in a CNN: the neuron outputs are discarded using saliency map as a distribution function. In [MOBPGVRA19a], we have experienced this new saliency-guided dropout scheme with AlexNet architecture on the Mexculture Dataset. Already comparing regular dropout on the basis of uniform distribution with saliency-based spatial dropout between last convolutional layer and the first FC layer, we get an increase of accuracy on the test set from 73.74% up to 79.13%. Hence saliency-based spatial dropout regularizes the model better than a simple dropout. This result is expected as we propagate salient information when blocking a non-salient one. When combining saliency-based spatial dropout with regular dropout in FC layers of AlexNet, we get even better results. The accuracy on the test set is 84.86%.

8.3.4.3 Use of Saliency in Backward Propagation

Dropping and pooling mechanisms are designed to process features only during forward propagation in deep neural networks. Then, once the output loss is computed for a given batch of images, the optimization process begins in the same way as in any other artificial neural network. But we have not exhausted all the potential of the information about relevant pixels/features. We can also use it when training synaptic weights, that is in the backward pass. During forward propagation we select relevant features randomly sampling accordingly the distribution defined by saliency map S_m. If we consider the response from the Eq. (8.14) and the saliency map values to

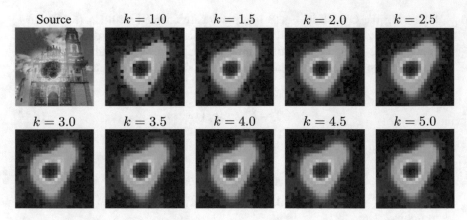

Fig. 8.7 Adaptive spatial saliency dropping. Visual attention maps are displayed in order to verify where activations are dropped. We stated that $k = 4$ brings better results after training several models. For this sample $\mu_{sm} = 0.22$ and λ^* is computed by Eq. (8.13) for each value of k

generate a binary map, with ones where $R(s) < S_m(x, y)$, then we can use it in the backward propagation in gradient descent. In the backward propagation the goal is to restrict the propagation of non-relevant local gradients. In the optimization process we use accelerated gradient descent with Nesterov's momentum [Nes83], see Chap. 4. The introduction of the saliency into backpropagation is realized by a simple element-wise multiplication of the gradients by the binary maps of each layer of the network:

$$\mathbf{V}_{i+1}^l = \mu \mathbf{V}_i^l - \eta \mathbf{R}_s^l \nabla \mathcal{L}(\mathbf{W}_i^l + \mu \mathbf{V}_i^l) \qquad (8.16)$$

$$\mathbf{W}_{i+1}^l = \mathbf{W}_i^l + \mathbf{V}_{i+1}^l$$

where $\mathbf{W}_i^l$ denotes the filters coefficients at iteration i, η the learning rate, μ the momentum and $\mathbf{V}_i^l$ is the velocity and $\mathbf{R}_s^l$ is the binary map generated by selection process at each layer. Once more to test the concept we have used the lightweight AlexNet architecture [MOBPGVRA19b]. Introduction of saliency maps into weights optimization in backward pass increased the accuracies, but only slightly. Indeed on the validation set of Mexculture dataset for recognition of 67 categories of buildings, the accuracy was 79.17% for full forward (saliency based pooling) and backward propagation vs 78.37% in case of only forward propagation with saliency-based pooling. Pretty similar behaviour is observed on test set: 76.54% for forward-backward vs 75.51% for forward only.

8.4 Conclusion

In this chapter we have focused on the application of Deep CNNs to the recognition of cultural content. But a simple application would not be attractive for our reader. We have reported on our experience in a very popular and exciting aspects of Deep NNs such as attention mechanisms. We have introduced visual attention models into the whole process of abstraction and classification ensured by these efficient classifiers. Indeed visual attention models when propagated through the CNNs perform better than new attention mechanisms such as Squeeze and Excitation networks and Double attention blocks. These last tools use the features highlighted by network itself to reinforce feature maps through the training and classification process. Visual attention we propose bring External information as it can be guided by top-down visual task, which is more semantic than a free observation of visual content. Such "importance" maps could be derived from other considerations, depending on application.

Chapter 9
Introducing Domain Knowledge

In this chapter we will consider another application case of Deep learning: classification of brain images for detection of Alzheimer's disease. In this particular application of medical imaging domain, Deep NNs have become the mandatory tool. In this chapter we give some highlights on how the usual steps in design of a Deep Neural Network classifier are implemented in the case when domain knowledge has to be considered. But more than that: faithful to our principle of showing new aspects of Deep Learning even in application cases, we will show how information fusion with siamese CNNs helps in increasing of performances of these classifiers.

9.1 Introduction

Introduction of image analysis and mining methods, e.g. classification into the vast domain of Computer Aided Diagnosis (CAD) has arrived very early. The famous Nagao's filter [NM79] was used for a non-linear enhancing of CT images already in 80th. The interest of Deep NNs for Alzheimer's disease detection in brain scans in different imaging modalities lies also in the fact that successful convolutional layers highlight brain structures, yielding finally quite decent classification scores. Nevertheless, the way to the application of CNNs for Alzheimer detection was quite long. And to understand it we will first focus on the classification problem to solve.

Alzheimer's Disease (AD) is one of the most frequent forms of dementia which progresses in developed countries due to the general aging of population. AD is a progressive neuro-degenerative disease accompanied by severe deterioration of human cognitive function and by—specifically—short-term memory loss. In order to develop non-invasive diagnostics methods, classification of brain scans of different modalities has become an intensively researched topic [LKB+17].

If we take an instantaneous picture of a screened cohort of subjects, then three clinical stages of AD can be distinguished:

© The Author(s), under exclusive license to Springer Nature Switzerland AG 2020
A. Zemmari, J. Benois-Pineau, *Deep Learning in Mining of Visual Content*,
SpringerBriefs in Computer Science, https://doi.org/10.1007/978-3-030-34376-7_9

Preclinical AD   About half of the people in this phase do not report cognitive troubles some years before AD diagnosis could be established, as cells' degeneration associated with AD begins years even decades before subjects would show clinical symptoms. Thus, an accurate diagnosis of the disease in the clinical phase is not yet possible.

Mild Cognitive Impairment (MCI)   Most patients go through the transitional stage called Mild Cognitive Impairment before they lapse into AD. At this stage of disease, a subject may show memory problems long before he gets an Alzheimer's diagnosis. Disease in this phase is not severe enough to disrupt a person's life. MCI is a challenging and confused group because in this phase the subject is not yet considered to have AD. Despite its large heterogeneity, MCI remains a group of interest in the study of early-stage AD.

Clinically Diagnosed AD   The late stage of Alzheimer's disease may also be called "severe". Subjects in this stage show decreased mental ability, total loss of cognitive function and finally this causes the death.

Current medical research studies are focusing on prediction of conversion or not of MCI cases into AD [LIK+18], but the classification problem using imaging modalities of the above mentioned three stages of the disease is far from being solved. The use of Deep CNNs for this task has drastically improved the classification scores, but the way to design the classification schemes, to focus on specific regions in the brain [LLL17] or on the whole brain[LCW+18] still remains a subject of research. In this chapter we will discuss how the domain—medical— knowledge can be used for design of Deep NN classifiers, selection of data for them, and the most complete use of different sources of information available for the same subjects due to screening with different imaging modalities.

9.2   Domain Knowledge

To design efficient algorithms for classification of brain images on the basis of CNN principles in the problem of AD diagnostics, we first need to introduce domain knowledge consisting in

- choice of image acquisition modalities
- choice of the areas in the brain affected by AD in priority
- choice of the voxels in the brain to feed into a CNN architecture

9.2.1   Imaging Modalities

Amongst different imaging modalities which have been successfully developed for medical imaging, Structural Magnetic Resonance Imaging (sMRI) [BCH+14] is the most commonly used for brain imaging for the purpose of AD detection as it provides information to describe the shape, the size and the integrity of gray and white matter structures in the brain [SILH$^+$16]. For a long time, the structural MRI is the most frequent technique for examining the anatomy and pathology of the brain, it is a widely used imaging technique in research as well as in clinical practice. In AD diagnostics, High-resolution T1-weighted sequences are used to distinct the anatomical boundaries and viewing structural changes in the brain [FFJJ$^+$10].

Diffusion Tensor Imaging (DTI) is another modality used in AD studies, it is a recent imaging technique that is able to track and quantify water diffusion along fiber bundles, to detect and describe the anisotropy on surrounding tissue micro-structure [BML94]. Three eigenvalues (λ_1, λ_2 and λ_3) and eigenvectors are calculated from water's molecular motion in a three-dimensional space to represent the main diffusion directions [BML94]. Fractional anisotropy (FA) and mean diffusivity (MD) are the two most common measures (scalar maps) derived from the diffusion tensor imaging.

In case of brain degeneration related to AD, the cerebro-spinal fluid fills-in the cavities and these effects are perceived both in FA and MD maps as the motion of water molecules becomes chaotic.

- *Mean diffusivity*: Mean diffusivity represents the average magnitude of a tensor's water diffusion and is equal to the average of the three eigenvalues ($\lambda_1 + \lambda_2 + \lambda_3$)/3. Mean diffusivity is the mean molecular motion in a certain voxel, but it provides no elements regarding diffusion directionality.

$$\text{MD} = \tilde{\lambda} = \frac{\lambda_1 + \lambda_2 + \lambda_3}{3} \tag{9.1}$$

- *Fractional anisotropy*: Fractional anisotropy (FA), a measure of the degree of diffusion anisotropy, is calculated from the standard formula:

$$\text{FA} = \sqrt{\frac{3}{2}} \sqrt{\frac{(\lambda_1 - \tilde{\lambda})^2 + (\lambda_2 - \tilde{\lambda})^2 + (\lambda_3 - \tilde{\lambda})^2}{\sqrt{\lambda_1^2 + \lambda_2^2 + \lambda_3^2}}} \tag{9.2}$$

Where $\tilde{\lambda}$ is the mean diffusivity (MD), or average rate of diffusion in all directions. Axial diffusivity is defined as the primary (largest) eigenvalue (AxD = λ_1), and captures the longitudinal diffusivity, or the diffusivity parallel to axonal fibers (assuming of course that the principal eigenvector is indeed following the dominant fiber direction, which may be unclear in regions with extensive fiber crossing). Radial diffusivity (RD), which captures the average diffusivity perpendicular to axonal fibers, is calculated as the average of the two smaller eigenvalues:

$$\text{RD} = \frac{\lambda_2 + \lambda_3}{2} \tag{9.3}$$

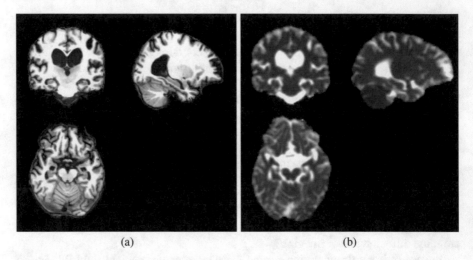

(a)   (b)

Fig. 9.1 Example of sMRI and DTI modalities on AD brain. (**a**) sMRI modality, (**b**) MD-DTI modality

An example of sMRI and MD-DTI modalities for the same brain of AD patient from ADNI database [JJBF+08] is presented in Fig. 9.1. We can see that both modalities reflect the same brain structure, but MD-DTI is the opposite compared to sMRI and sMRI is more contrasted.

9.2.2 Selection of Brain ROI

When a person is suffering from AD, some regions-of-interest (ROIs) in the brain are known to be affected by AD such as the Hippocampus. The latter is to be affected first [DLP+95, DGS+01]. Temporal and cingulate gyri, and precuneus are amongst other affected structures [BB91, JPX+99]. Hippocampal degeneration, i.e. shrinkage remains the first visible degradation of brain structures. Therefore, it is justified to focus classification of brain scans on this ROI in priority. To illustrate the interest of using different modalities for classification task in AD detection, we present, in Fig. 9.2 an example of scans of three subjects from ADNI database. The upper row contains from left-to-right sMRI slices of a sagittal projection of AD, MCI and healthy brain and the lower row contains MD-DTI of the same projection. It can be seen that the degeneration of hippocampus is expressed by a prevalence of dark pixels corresponding to Cerebro-Spinal Fluid (CSF) in sMRI and of white pixels, corresponding to a chaotic motion of water molecules in CSF in the same region in MD-DTI. To select the ROI in a brain scan of a subject, the brain atlas AAL [TMLP+02] is used. In this atlas all voxels belonging to the same brain structure, e.g. hippocampus have the same label. But before the hippocampal voxels could be

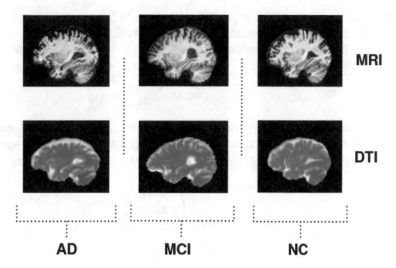

Fig. 9.2 Examples of AD, MCI and Healthy brain on SMRI and DTI Modalities. From left to right AD, MCI, NC

selected, we need to a align the brain scan of a subject to the so-called MNI template [FTS+05] for which the AAL was built.

9.2.3 Alignment of Different Imaging Modalities

In order to use both modalities sMRI and MD-DTI in classification of subjects, we need to register them in the same space. Hence the same physical parts of the brain will be analysed. Thus we align the sMRI scan of the subject on MNI template and then co-register the DTI modality to corresponding aligned MRI as illustrated in Fig. 9.4.

For the alignment on MNI: as we look for preserving the pattern of the ROI and avoiding features deformation, both image modalities are registered using the 3D affine transformation [ANC+97]. This step of pre-processing includes translation, rotation, scaling and shearing operations [AF97]. The main goal here is to fit a given image (f) to a template image (g), using 12-parameter affine transformation (q_1 to q_{12}).

From the position $x = (x_1, x_2, x_3)^\mathsf{T}$ in the image f to the position $y = (y_1, y_2, y_3)^\mathsf{T}$ in g, we can define the affine transformation mapping as follows:

$$
\begin{pmatrix} y_1 \\ y_2 \\ y_3 \\ 1 \end{pmatrix} = \begin{pmatrix} m_1 & m_4 & m_7 & m_{10} \\ m_2 & m_5 & m_8 & m_{11} \\ m_3 & m_6 & m_9 & m_{12} \\ 0 & 0 & 0 & 1 \end{pmatrix} \begin{pmatrix} x_1 \\ x_2 \\ x_3 \\ 1 \end{pmatrix}
\tag{9.4}
$$

we refer this mapping equation as $\mathbf{y} = \mathbf{M} \times \mathbf{x}$, where $\mathbf{M}$ is the mapping matrix and m_i elements are a function of parameters q_1 to q_{12}, see Eqs. (9.6)–(9.8) below.

The matrix $\mathbf{M}$ can be decomposed as product of four matrices: translation, rotation, scaling and shearing.

$$\mathbf{M} = M_{Translation} \times M_{Rotation} \times M_{Scaling} \times M_{Shearing} \tag{9.5}$$

The parameters q_1, q_2, and q_3 correspond to three translation parameters, q_4, q_5, and q_6 correspond to three rotation parameters. q_7, q_8, and q_9 to three zooms and finally q_{10}, q_{11}, and q_{12} are the three shear parameters.

$$M_{Translation} = \begin{pmatrix} 1 & 0 & 0 & q_1 \\ 0 & 1 & 0 & q_2 \\ 0 & 0 & 1 & q_3 \\ 0 & 0 & 0 & 1 \end{pmatrix} \tag{9.6}$$

$$M_{Rotation} = \begin{pmatrix} 1 & 0 & 0 & 0 \\ 0 & \cos(q_4) & \sin(q_4) & 0 \\ 0 & -\sin(q_4) & \cos(q_4) & 0 \\ 0 & 0 & 0 & 1 \end{pmatrix} \times \begin{pmatrix} \cos(q_5) & 0 & \sin(q_5) & 0 \\ 0 & 0 & 0 & 0 \\ -\sin(q_5) & 0 & \cos(q_5) & 0 \\ 0 & 0 & 0 & 1 \end{pmatrix} \times \begin{pmatrix} \cos(q_6) & \sin(q_6) & 0 & 0 \\ -\sin(q_6) & \cos(q_6) & 0 & 0 \\ 0 & 0 & 1 & 0 \\ 0 & 0 & 0 & 1 \end{pmatrix}$$

$$M_{Scaling} = \begin{pmatrix} q_7 & 0 & 0 & 0 \\ 0 & q_8 & 0 & 0 \\ 0 & 0 & q_9 & 0 \\ 0 & 0 & 0 & 1 \end{pmatrix} \tag{9.7}$$

$$M_{Shearing} = \begin{pmatrix} 1 & q_{10} & q_{11} & 0 \\ 0 & 1 & q_{12} & 0 \\ 0 & 0 & 1 & 0 \\ 0 & 0 & 0 & 1 \end{pmatrix} \tag{9.8}$$

The parameter estimation is realized by minimizing the sum of squared differences (SSD) objective between the object (f) and template image (g) using the Gauss Newton algorithm [FAF$^+$95]. The images may be scaled differently, so an additional parameter (w) is needed to accommodate this difference. The function to minimize is then:

$$SSD(f, g) = \sum_{i=1}^{I} (f(\underbrace{M * x_i}_{y_i}) - wg(x_i))^2 \tag{9.9}$$

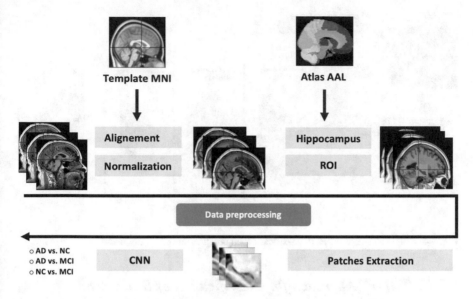

Fig. 9.3 Diagram of the Preprocessing for sMRI dataset

The template to which each T1-weighted anatomical (sMRI) scan is aligned, is the so-called MNI template produced at the Montreal Neurological Institute (MNI) [FTS$^+$05]. It is build from 152 normal subjects by averaging scans. The alignment on the MNI template is illustrated in Fig. 9.3 [ABBP$^+$17]. The next step in data preparation is data normalization. Here voxel intensity is normalized for the whole dataset in order to have similar intensities for similar structures. The process is done using the software SPM8 to fulfill the registration [Fri96].

The next step is the co-registration of MD-DTI modality to sMRI. Here the estimated transformation matrix M is applied to the MD-DTI scan to align it to MNI as well. A preliminary step is the skull-stripping in order to remove the bright skull voxels.

9.3 Siamese Deep NNs for Fusion of Modalities

In both modalities the characteristic features of our classification problem are noticeable: it is a deformation of hippocampal ROI. Hence it is very much interesting to make the modalities collaborate in the same Deep learning framework to reinforce the performances of classifier. Indeed, today it is clear that information fusion is the sure way to enhance performances of classifiers [IBPQ14]. But before we will present our classification approach for one modality. We call it "2D+ϵ" approach, as we do not use a full 3D information, but only a limited number of slices in a 3D volume of brain scans.

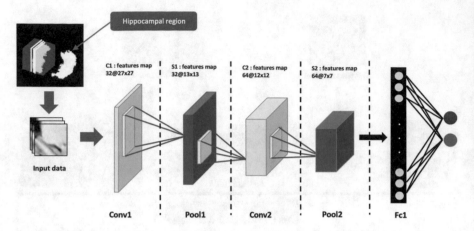

Fig. 9.4 Single Network for ROI classification on one projection of one modality

9.3.1   "2D+ε" Approach for Classification of Brain Scans for AD Detection

The 2D+ε approach was proposed in [ABBP+17] for brain ROI classification. This means that 2D convolutions are used in a CNN architecture feeding it with three neighbouring slices of the ROI volume extracted from one of three imaging projections of the brain: Sagittal, Axial, and Coronal whatever is the modality: sMRI or MD-DTI. The median slice of Hippocampal ROI and its two neighbours have been selected to feed the network, see Fig. 9.4.

The network architecture we designed is relatively shallow. Indeed it is conditioned by the low resolution of our ROI - hippocampal area which is of $28 \times 28 \times 28$ voxels. Hence in a 2D+ε setting, the extracted brain volume is of $28 \times 28 \times 3$. We consider it as a three-channel 2D volume of size 28×28.

The inspiration of proposed architecture, illustrated in Fig. 9.4 comes from LeNet presented in Chap. 6. It consists of two convolutional layers followed by max pooling layer for each one, and one fully connected layer.

The classification task is binary. Indeed in literature devoted to the diagnostics of AD on brain scans the two-class formulation AD/NC, AD/MCI, MCI/NC is reported as more efficient [YTQ12].

Working on a single modality (sMRI) we discovered that the sagittal projection was the most discriminant. Indeed in [ABPAC17] experiments conducted on ADNI-1 dataset of 815 images with domain specific data augmentation by translating the 3D ROI in the brain volume yielded the following accuracies. For AD/NC in sagittal projection we got 82.80%, while in coronal it was 80.15% and in axial 79.69%. The same trend was observed for other binary classification tasks with the most difficult separation of MCI/NC classes with accuracy of 66.12% on sagittal, 57.56% on coronal and 61.25% on axial. Fusion of single projection results was realized by a majority vote operator on classification scores, which yielded for AD/NC 91.41%, AD/MCI 69.53%, and MCI/NC 65.62%.

9.3.2   Siamese Architecture

With the introduction of siamese NNs [BBB$^+$93, BGL$^+$94], the so-called "intermediate fusion" we practice a lot in mining of complex visual information [PKL$^+$12] becomes possible in the formalism of CNNs. Hence, Kosh et al. [KZS15] proposed siamese CNNs for one shot image recognition. They use a siamese network for learning similarity between two images. Each image is submitted to one siamese twin, i.e. one branch of a convolutional network. The difference between the vectors $\mathbf{h}_{1,L-1}^{(j)}$ and $\mathbf{h}_{2,L-1}^{(j)}$ from fully connected layers of both branches 1 and 2 is used to compute the prediction $\mathbf{P}$:

$$\mathbf{P} = \sigma(\sum_j \alpha_j \left| \mathbf{h}_{1,L-1}^{(j)} - \mathbf{h}_{2,L-1}^{(j)} \right|) \qquad (9.10)$$

Here σ is the sigmoidal activation function, and α_j are supplementary parameters learned by the algorithm during training.

In our case, the siamese architecture is designed for fusion of information. Here the concatenation of features in FC layer is realized linking six networks of both modalities and all projections in each of them. We thus build a full siamese architecture presented in Fig. 9.5. Here from left-to-right we have the input of three slices of each projection of each modality. Then the single branch network is designed and parameterized as that one presented in Sect. 9.3.1 and illustrated in Fig. 9.4. Finally the fusion layer consists in concatenation of features from six fully connected layers. Here once more we have a set of binary classification problems AD/NC, AD/MCI, MCI/NC and with two modalities available for each subject.

The results of such a fusion of modalities were obtained on a mix of ADNI-2, ADNI-GO,ADNI-3 database with 736 images in overall. As in many screened cohorts the distribution of subjects data between three classes was quite unequal. Thus data of 390 NC subjects are available, for MCI the amount is lower—273 scans and only 64 sMRI and DTI scans are available for patients with Alzheimer's disease. All these data were split into 528 training samples, 148 validation samples and 60 test samples. Then, after domain-dependent data augmentation by random translations of bounding cuboid inside the brain volume and Gaussian blurring as in [KABP$^+$18, AKK$^+$18] the total amount of data for each dataset was: 89,700 for training dataset, 24,000 for validation dataset, 600 for test dataset. The networks were trained from scratch by stochastic gradient descent with Nesterov momentum (see Chap. 4). The hypothesis that fusion of modalities outperforms single modality was confirmed experimentally. Indeed, on the most discriminative—sMRI modality

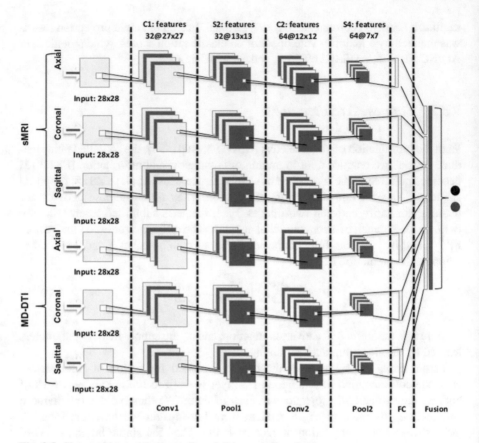

Fig. 9.5 Siamese Network built on all modalities and all projections

only, the following results were obtained on the presented database using the most discriminative sagittal projection: AD/NC 82.92%, AD/MCI 66.73%, and MCI/NC 65.51%.

With the full siamese network we got increased accuracies: for AD/NC 86.45%, AD/MCI 72.83%, and MCI/NC 70.59%.

9.4   Conclusion

Hence in this chapter we have shown a case-study of design of Deep CNN architectures for a specific and highly researched domain: classification of medical images for computer-aided diagnostics of Alzheimer disease. As the set of methods proposed nowadays in the framework of Deep learning formalism, is very rich, we took profit of this to introduce another aspect: information fusion with Deep CNNs. In case of medical images the domain knowledge has to be introduced in the whole process of the design of classifiers. First of all, the data used in fusion framework have to be properly prepared and aligned. We have shown this on sMRI and MD-DTI modalities. Then, medical knowledge has to be explored in order to select meaningful areas—ROIs—in the images to submit to classifiers and to avoid heavy computations on the whole image. Furthermore, the relatively low resolution of medical images from sMRI, DTI and other modalities prevent from design of very deep convolutional networks. We have obviously limit their depth if using the standard duo of operations: $n \times n$ convolutions and pooling to abstract features from layer-to-layer. Finally, the limited cardinals of image databases make us to develop domain-dependent data augmentation techniques. Last, but not least, in this chapter we have introduced siamese networks and have showed how this architectures allow for increase of classification performances when using data from different modalities describing the same phenomenon.

Conclusion

Hence in this book we have focused on theoretical aspects of the winning model Deep learning.

In Chap. 2 we have started with the general supervised learning problem formulation and presented evaluation metrics.

In Chap. 3 the fundamental model of Neural Networks which is the basis of Deep learning was analyzed.

In Chap. 4 we have introduced popular optimization methods used for training of parameters for Deep NNs.

In Chap. 5 coming to our primary task of visual information mining we tried to trace an analogy between two operations any convolution neural network fulfills today: convolution and sampling with the use of these operations in general Image Processing and analysis approaches.

Chapter 6 was devoted to the detailed explanation of CNNs—which are used for both static image and video understanding when we are working on an independent frame-by-frame basis.

In Chap. 7 we were interested in temporal analysis of visual content, we first made a flash back to a very popular model such as HMMs which has been used for recognition of temporal events in video and we bridged them to the RNNs and LSTMs—more modern way to do this.

Having conducted research in visual information mining with these models we were eager to share it with application examples in mining of digital cultural content and in Computer Aided Diagnostics in Chaps. 8 and 9. Here we have more practically introduced attention mechanisms and seamese networks. Concluding this book we hope that it can serve as a pocket hand book for actual generation of young researches and professionals.

A. Zemmari, J. Benois-Pineau, *Deep Learning in Mining of Visual Content*,
SpringerBriefs in Computer Science, https://doi.org/10.1007/978-3-030-34376-7

Glossary

This glossary introduces notation definitions used in this book. It can happen that the notation changes in some sections. In this case, the new notation will be clearly explained.

a	A scalar.
$\mathbf{a}$	A vector.
η	Learning rate.
$f(x; \boldsymbol{\theta})$	A function parametrized with $\boldsymbol{\theta}$, applied to a variable x.
$\mathcal{L}(.)$	Loss function.
$\nabla f(.)$	Gradient vector of function f.
M^T	Transpose of matrix M.
N	Size of the dataset.
$\frac{\partial F}{\partial x}$	Partial derivative of F with respect to x.
(x, y)	An example from the dataset: x is a vector of features and y is the corresponding label.

© The Author(s), under exclusive license to Springer Nature Switzerland AG 2020   101
A. Zemmari, J. Benois-Pineau, *Deep Learning in Mining of Visual Content*,
SpringerBriefs in Computer Science, https://doi.org/10.1007/978-3-030-34376-7

References

ABBP+17. Karim Aderghal, Manuel Boissenin, Jenny Benois-Pineau, Gwenaëlle Catheline, and Karim Afdel. Classification of sMRI for AD diagnosis with convolutional neuronal networks: A pilot 2d+ϵ study on ADNI. In *International Conference on Multimedia Modeling*, pages 690–701. Springer, Cham, 2017.

ABPAC17. Karim Aderghal, Jenny Benois-Pineau, Karim Afdel, and Gwenaëlle Catheline. FuseMe: Classification of sMRI images by fusion of deep CNNs in 2d+ϵ projections. In *Proceedings of the 15th International Workshop on Content-Based Indexing*, page 34. ACM, 2017.

AF97. John Ashburner and K Friston. Multimodal image coregistration and partitioning—a unified framework. *Neuroimage*, 6(3):209–217, 1997.

AKK+18. Karim Aderghal, Alexander Khvostikov, Andrei Krylov, Jenny Benois-Pineau, Karim Afdel, and Gwenaelle Catheline. Classification of Alzheimer disease on imaging modalities with deep CNNs using cross-modal transfer learning. In *2018 IEEE 31st International Symposium on Computer-Based Medical Systems (CBMS)*, pages 345–350. IEEE, 2018.

ANC+97. John Ashburner, P Neelin, DL Collins, A Evans, and K Friston. Incorporating prior knowledge into image registration. *Neuroimage*, 6(4):344–352, 1997.

ASJ+07. Haider Ali, Christin Seifert, Nitin Jindal, Lucas Paletta, and Gerhard Paar. Window detection in facades. In *Image Analysis and Processing, 2007. ICIAP 2007. 14th International Conference on*, pages 837–842. IEEE, 2007.

BABPA+17. Olfa Ben Ahmed, Jenny Benois-Pineau, Michelle Allard, Gwénaëlle Catheline, Chokri Ben Amar, Alzheimer's Disease Neuroimaging Initiative, et al. Recognition of Alzheimer's disease and mild cognitive impairment with multimodal image-derived biomarkers and multiple kernel learning. *Neurocomputing*, 220:98–110, 2017.

BB91. Heiko Braak and Eva Braak. Neuropathological staging of Alzheimer-related changes. *Acta neuropathologica*, 82(4):239–259, 1991.

BBB+93. Jane Bromley, James W. Bentz, Léon Bottou, Isabelle Guyon, Yann LeCun, Cliff Moore, Eduard Säckinger, and Roopak Shah. Signature verification using A "siamese" time delay neural network. *IJPRAI*, 7(4):669–688, 1993.

BCH+14. Robert W. Brown, Y.-C. Norman Cheng, Mark E. Haacke, Michael R. Thomson, and Ramesh Vnekatesan. *Magnetic Resonance Imaging: Physical Principles and Sequence Design 2nd Edition*. John Wiley & Sons, Inc. N.Y., 2014.

BETVG08. Herbert Bay, Andreas Ess, Tinne Tuytelaars, and Luc Van Gool. Speeded-up robust features (surf). *Computer vision and image understanding*, 110(3):346–359, 2008.

BGL+94. Jane Bromley, Isabelle Guyon, Yann LeCun, Eduard Säckinger, and Roopak Shah. Signature verification using a "siamese" time delay neural network. In *Advances in neural information processing systems*, pages 737–744, 1994.

BGM07. Alexander C Berg, Floraine Grabler, and Jitendra Malik. Parsing images of architectural scenes. In *Computer Vision, 2007. ICCV 2007. IEEE 11th International Conference on*, pages 1–8. IEEE, 2007.

BML94. Peter J Basser, James Mattiello, and Denis LeBihan. MR diffusion tensor spectroscopy and imaging. *Biophysical journal*, 66(1):259–267, 1994.

BPC17. Jenny Benois-Pineau and Patrick Le Callet, editors. *Visual Content Indexing and Retrieval with Psychovisual models*. Springer, Heidelberg, New York, Dordrecht, London, 2017.

BPPC12. Jenny Benois-Pineau, Frdric Precioso, and Matthieu Cord. *Visual Indexing and Retrieval*. Springer Publishing Company, Incorporated, 2012.

BPSW70. Leonard E. Baum, Ted Petrie, George Soules, and Norman Weiss. A maximization technique occurring in the statistical analysis of probabilistic functions of Markov chains. *Ann. Math. Statist.*, 41(1):164–171, 02 1970.

BV92. Gyon Isabelle M. Boser, Bernhard E. and Vladimir N. Vapnik. A training algorithm for optimal margin classifiers. In *COLT '92 Proceedings of the fifth annual workshop on Computational learning theory*, pages 144–152. ACM, 1992.

ByFL99. S. Ben-Yacoub, B. Fasel, and J. Lüttin. Fast face detection using MLP and FFT. In *in Proc. Second International Conference on Audio and Video-based Biometric Person Authentication (AVBPA'99*, pages 31–36, 1999.

CDF+04. Gabriella Csurka, Christopher R. Dance, Lixin Fan, Jutta Willamowski, and Cédric Bray. Visual categorization with bags of keypoints. In *In Workshop on Statistical Learning in Computer Vision, ECCV*, pages 1–22, 2004.

CH+67. Thomas M Cover, Peter Hart, et al. Nearest neighbor pattern classification. *IEEE transactions on information theory*, 13(1):21–27, 1967.

Chu14. Wei-Ta Chu. An introduction to optimization. 2014.

CKL+18. Yunpeng Chen, Yannis Kalantidis, Jianshu Li, Shuicheng Yan, and Jiashi Feng. A^2-nets: Double attention networks. In S. Bengio, H. Wallach, H. Larochelle, K. Grauman, N. Cesa-Bianchi, and R. Garnett, editors, *Advances in Neural Information Processing Systems 31*, pages 352–361. Curran Associates, Inc., 2018.

CV95. Corinna Cortes and Vladimir Vapnik. Support-vector networks. *Machine learning*, 20(3):273–297, 1995.

CVMBB14. Kyunghyun Cho, Bart Van Merriënboer, Dzmitry Bahdanau, and Yoshua Bengio. On the properties of neural machine translation: Encoder-decoder approaches. *arXiv preprint arXiv:1409.1259*, 2014.

Cyb89. George Cybenko. Approximation by superpositions of a sigmoidal function. *MCSS*, 2(4):303–314, 1989.

DBL13. *Proceedings of the 30th International Conference on Machine Learning, ICML 2013, Atlanta, GA, USA, 16–21 June 2013*, volume 28 of *JMLR Workshop and Conference Proceedings*. JMLR.org, 2013.

DGS+01. Bradford C Dickerson, I Goncharova, MP Sullivan, C Forchetti, RS Wilson, DA Bennett, Laurel A Beckett, and L deToledo Morrell. MRI-derived entorhinal and hippocampal atrophy in incipient and very mild Alzheimer's disease. *Neurobiology of aging*, 22(5):747–754, 2001.

DHS11. John Duchi, Elad Hazan, and Yoram Singer. Adaptive subgradient methods for online learning and stochastic optimization. 12:2121–2159, July 2011. http://jmlr.org/papers/volume12/duchi11a/duchi11a.pdf.

DLP⁺95.   Bernard Deweer, Stephane Lehericy, Bernard Pillon, Michel Baulac, J Chiras, C Marsault, Y Agid, and B Dubois. Memory disorders in probable Alzheimer's disease: the role of hippocampal atrophy as shown with MRI. *Journal of Neurology, Neurosurgery & Psychiatry*, 58(5):590–597, 1995.

DLR77.   A. P. Dempster, N. M. Laird, and D. B. Rubin. Maximum likelihood from incomplete data via the EM algorithm. *JOURNAL OF THE ROYAL STATISTICAL SOCIETY, SERIES B*, 39(1):1–38, 1977.

DPG⁺14.   Yann Dauphin, Razvan Pascanu, Çaglar Gülçehre, Kyunghyun Cho, Surya Ganguli, and Yoshua Bengio. Identifying and attacking the saddle point problem in high-dimensional non-convex optimization. *CoRR*, abs/1406.2572, 2014. http://arxiv.org/abs/1406.2572.

Duc07.   Andrew T Duchowski. Eye tracking methodology. *Theory and practice*, 328, 2007.

FAF⁺95.   Karl J Friston, John Ashburner, Christopher D Frith, J-B Poline, John D Heather, and Richard SJ Frackowiak. Spatial registration and normalization of images. *Human brain mapping*, 3(3):165–189, 1995.

Faw06.   Tom Fawcett. An introduction to roc analysis. *Pattern recognition letters*, 27(8):861–874, 2006.

FB81.   Martin A Fischler and Robert C Bolles. Random sample consensus: a paradigm for model fitting with applications to image analysis and automated cartography. *Communications of the ACM*, 24(6):381–395, 1981.

FFJJ⁺10. Giovanni B Frisoni, Nick C Fox, Clifford R Jack Jr, Philip Scheltens, and Paul M Thompson. The clinical use of structural MRI in Alzheimer disease. *Nature Reviews Neurology*, 6(2):67, 2010.

Fli04.   Flickr. Flickr: Find your inspiration, 2004.

Fri96.   Karl J Friston. Statistical parametric mapping and other analyses of functional imaging data. *Brain mapping: The methods*, 1996.

FSM⁺17.   N Iandola Forrest, Han Song, W Matthew, Ashraf Khalid, and J William Dally. SqueezeNet: AlexNet-level accuracy with 50x fewer parameters and< 0.5 MB model size. pages 207–212, 2017.

FTS⁺05.   GB Frisoni, C Testa, F Sabattoli, A Beltramello, H Soininen, and MP Laakso. Structural correlates of early and late onset Alzheimer's disease: voxel based morphometric study. *Journal of Neurology, Neurosurgery & Psychiatry*, 76(1):112–114, 2005.

Fu82.   K.S. Fu. *Syntactic pattern recognition and applications*. Prentice Hall, 1982.

GBCB16.   Ian Goodfellow, Yoshua Bengio, Aaron Courville, and Yoshua Bengio. *Deep learning*, volume 1. MIT Press Cambridge, 2016.

GDBBP16. Iván González-Díaz, Vincent Buso, and Jenny Benois-Pineau. Perceptual modeling in the problem of active object recognition in visual scenes. *Pattern Recognition*, 56:129–141, 2016.

GDDM13.   Ross B. Girshick, Jeff Donahue, Trevor Darrell, and Jitendra Malik. Rich feature hierarchies for accurate object detection and semantic segmentation. *CoRR*, abs/1311.2524, 2013. http://arxiv.org/abs/1311.2524.

Gir15.   Ross B. Girshick. Fast R-CNN. *CoRR*, abs/1504.08083, 2015. http://arxiv.org/abs/1504.08083.

Hau07.   Raphael Hauser. Line search methods for unconstrained optimisation. *Lecture 8, Numerical Linear Algebra and Optimisation Oxford University Computing Laboratory*, 2007. https://people.maths.ox.ac.uk/hauser/hauser_lecture2.pdf.

HHP01.   Bernd Heisele, Purdy Ho, and Tomaso A. Poggio. Face recognition with support vector machines: Global versus component-based approach. In *ICCV*, pages 688–694. IEEE Computer Society, 2001.

HKP07.   Jonathan Harel, Christof Koch, and Pietro Perona. Graph-based visual saliency. In *Advances in neural information processing systems*, pages 545–552, 2007.

HS97.   Sepp Hochreiter and Jürgen Schmidhuber. Long short-term memory. *Neural Computation*, 9(8):1735–1780, 1997.

HSS12.   Geoffrey Hinton, Nitish Srivastava, and Kevin Swersky. Neural networks for machine learning - lecture 6a - overview of mini-batch gradient descent. 2012. http://www.cs.toronto.edu/~tijmen/csc321/slides/lecture_slides_lec6.pdf.

HW68.   D. H. HUBEL and T. N. WIESEL.   Receptive fields and functional architecture of monkey striate cortex.   195:215–243, 1968. http://hubel.med.harvard.edu/papers/HubelWiesel1968Jphysiol.pdf.

HZRS15.   Kaiming He, Xiangyu Zhang, Shaoqing Ren, and Jian Sun. Deep residual learning for image recognition. *CoRR*, abs/1512.03385, 2015. http://arxiv.org/abs/1512.03385.

IBPQ14.   Bogdan Ionescu, Jenny Benois-Pineau, Tomas Piatrik, and Georges Quénot, editors. *Fusion in Computer Vision - Understanding Complex Visual Content*. Advances in Computer Vision and Pattern Recognition. Springer, 2014.

JH08.   Orlando De Jesus and Martin T. Hagan. Backpropagation through time for general dynamic networks. In Hamid R. Arabnia and Youngsong Mun, editors, *Proceedings of the 2008 International Conference on Artificial Intelligence, ICAI 2008, July 14–17, 2008, Las Vegas, Nevada, USA, 2 Volumes (includes the 2008 International Conference on Machine Learning; Models, Technologies and Applications)*, pages 45–51. CSREA Press, 2008.

JJBF⁺08. Clifford R Jack Jr, Matt A Bernstein, Nick C Fox, Paul Thompson, Gene Alexander, Danielle Harvey, Bret Borowski, Paula J Britson, Jennifer L. Whitwell, Chadwick Ward, et al. The Alzheimer's disease neuroimaging initiative (ADNI): MRI methods. *Journal of Magnetic Resonance Imaging: An Official Journal of the International Society for Magnetic Resonance in Medicine*, 27(4):685–691, 2008.

JLLT18.   Saumya Jetley, Nicholas A. Lord, Namhoon Lee, and Philip H. S. Torr. Learn to pay attention. *CoRR*, abs/1804.02391, 2018.

JPX⁺99.   Clifford R Jack, Ronald C Petersen, Yue Cheng Xu, Peter C O'Brien, Glenn E Smith, Robert J Ivnik, Bradley F Boeve, Stephen C Waring, Eric G Tangalos, and Emre Kokmen. Prediction of ad with MRI-based hippocampal volume in mild cognitive impairment. *Neurology*, 52(7):1397–1397, 1999.

KABP⁺18. Alexander Khvostikov, Karim Aderghal, Jenny Benois-Pineau, Andrey Krylov, and Gwenaelle Catheline.   3D CNN-based classification using sMRI and MD-DTI images for Alzheimer disease studies. *arXiv preprint arXiv:1801.05968*, 2018.

KB14.   Diederik P. Kingma and Jimmy Ba. Adam: A method for stochastic optimization. *CoRR*, abs/1412.6980, 2014. https://arxiv.org/pdf/1412.6980.pdf.

KBD⁺14.   Svebor Karaman, Jenny Benois-Pineau, Vladislavs Dovgalecs, Rémi Mégret, Julien Pinquier, Régine André-Obrecht, Yann Gaëstel, and Jean-François Dartigues. Hierarchical hidden Markov model in detecting activities of daily living in wearable videos for studies of dementia. *Multimedia Tools Appl.*, 69(3):743–771, 2014.

KOG03.   Ewa Kijak, Lionel Oisel, and Patrick Gros. Temporal structure analysis of broadcast tennis video using hidden Markov models. In Minerva M. Yeung, Rainer Lienhart, and Chung-Sheng Li, editors, *Storage and Retrieval for Media Databases 2003, Santa Clara, CA, USA, January 22, 2003*, volume 5021 of *SPIE Proceedings*, pages 289–299. SPIE, 2003.

KSH12.   Alex Krizhevsky, Ilya Sutskever, and Geoffrey E Hinton. Imagenet classification with deep convolutional neural networks. In *Advances in neural information processing systems*, pages 1097–1105, 2012.

KZS15.   Gregory Koch, Richard Zemel, and Ruslan Salakhutdinov. Siamese neural networks for one-shot image recognition. In *ICML deep learning workshop*, volume 2, 2015.

LBBH98. Yann Lecun, Léon Bottou, Yoshua Bengio, and Patrick Haffner. Gradient-based learning applied to document recognition. pages 2278–2324, 1998.

LBOM98. Yann LeCun, Leon Bottou, Genevieve B. Orr, and Klaus-Robert MÃijller. Efficient backprop. pages 9–50, 1998. http://yann.lecun.com/exdb/publis/pdf/lecun-98b.pdf.

LCW+18. Manhua Liu, Danni Cheng, Kundong Wang, Yaping Wang, Alzheimer's Disease Neuroimaging Initiative, et al. Multi-modality cascaded convolutional neural networks for Alzheimer's disease diagnosis. *Neuroinformatics*, pages 1–14, 2018.

LeC. Yann LeCun. MNIST Demos. *Yann LeCun's website*. http://yann.lecun.com/exdb/lenet/index.html.

LIK+18. Collin C. Luk, Abdullah Ishaque, Muhammad Khan, Daniel Ta, Sneha Chenji, Yee-Hong Yang, Dean Eurich, and Sanjay Kalra. Alzheimer's disease: 3-dimensional MRI texture for prediction of conversion from mild cognitive impairment. *Alzheimer's & Dementia: Diagnosis, Assessment & Disease Monitoring*, 10:755 – 763, 2018.

Lin94. Tony Lindeberg. *Scale-Space Theory in Computer Vision*, volume 256 of *The Springer International Series in Engineering and Computer Science*. Springer, 1994.

LKB+17. Geert Litjens, Thijs Kooi, Babak Ehteshami Bejnordi, Arnaud Arindra Adiyoso Setio, Francesco Ciompi, Mohsen Ghafoorian, Jeroen Awm Van Der Laak, Bram Van Ginneken, and Clara I Sánchez. A survey on deep learning in medical image analysis. *Medical image analysis*, 42:60–88, 2017.

LLL17. Suhuai Luo, Xuechen Li, and Jiaming Li. Automatic Alzheimer's disease recognition from MRI data using deep learning method. *Journal of Applied Mathematics and Physics*, 5(09):1892, 2017.

Low04. David G Lowe. Distinctive image features from scale-invariant keypoints. *International journal of computer vision*, 60(2):91–110, 2004.

LRM15. Tsung-Yu Lin, Aruni RoyChowdhury, and Subhransu Maji. Bilinear CNN models for fine-grained visual recognition. In *Proceedings of the IEEE international conference on computer vision*, pages 1449–1457, 2015.

Mac67. James MacQueen. Some methods for classification and analysis of multivariate observations. In *Proceedings of the fifth Berkeley symposium on mathematical statistics and probability*, volume 1, pages 281–297. Oakland, CA, USA, 1967.

MBL04. Francesca Manerba, Jenny Benois-Pineau, and Riccardo Leonardi. Extraction of foreground objects from an MPEG2 video stream in rough-indexing framework. In *Storage and Retrieval Methods and Applications for Multimedia*, volume 5307 of *SPIE Proceedings*, pages 50–60. SPIE, 2004.

McC43. Pitts W. McCulloch, W. S. A logical calculus of the ideas immanent in nervous activity. *Bulletin of Mathematical Biophysics*, 5:115–133, 1943.

Min87. Papert Seymour Minsky, Marvin. *Perceptrons. An Introduction to Computational Geometry, Expanded Edition*. MIT Press, 1987.

MJ98. Sheng Ma and Chuanyi Ji. A unified approach on fast training of feedforward and recurrent networks using EM algorithm. *IEEE Trans. Signal Processing*, 46(8):2270–2274, 1998.

MMW+11. Markus Mathias, Andelo Martinovic, Julien Weissenberg, Simon Haegler, and Luc Van Gool. Automatic architectural style recognition. *ISPRS-International Archives of the Photogrammetry, Remote Sensing and Spatial Information Sciences*, 3816:171–176, 2011.

MOBG⁺18. Abraham Montoya Obeso, Jenny Benois-Pineau, Kamel Guissous, Valérie Gouet-Brunet, Mireya Saraí García Vázquez, and Alejandro Alvaro Ramírez-Acosta. Comparative study of visual saliency maps in the problem of classification of architectural images with deep CNNs. In *2018 Eighth International Conference on Image Processing Theory, Tools and Applications (IPTA)*, pages 1–6. IEEE, 2018.

MOBPGVRA18. Abraham Montoya Obeso, Jenny Benois-Pineau, Mireya Saraí García Vázquez, and Alejandro Alvaro Ramírez Acosta. Introduction of Explicit Visual Saliency in Training of Deep CNNs: Application to Architectural Styles Classification. In *Proceedings of the 16th International Conference on Content-Based Multimedia Indexing*, page 16. IEEE, 2018.

MOBPGVRA19a. Abraham Montoya Obeso, Jenny Benois-Pineau, Mireya Saraí García Vázquez, and Alejandro Alvaro Ramírez Acosta. Dropping activations in convolutional neural networks with visual attention maps. In *Proceedings of the 17th International Conference on Content-Based Multimedia Indexing*, page 4, 2019.

MOBPGVRA19b. Abraham Montoya Obeso, Jenny Benois-Pineau, Mireya Saraí García Vázquez, and Alejandro Alvaro Ramírez Acosta. Forward-backward visual saliency propagation in deep NNs vs internal attentional mechanisms. In *2019 International Conference on Image Processing Theory, Tools and Applications (IPTA)*, pages 1–6. IEEE, 2019.

Nes83. Yurii Nesterov. A method of solving a convex programming problem with convergence rate $O(1/k^2)$. *Soviet Mathematics Doklady (Vol. 27)*, 1983.

Neu75. David L. Neuhoff. The Viterbi algorithm as an aid in text recognition (corresp.). *IEEE Trans. Information Theory*, 21(2):222–226, 1975.

NM79. M. Nagao and T. Matsyama. Edge preserving smoothing. *Computer graphics and image processing*, 9:394âĂŞ–407, 1979.

ORR⁺16. Abraham Montoya Obeso, Laura Mariel Amaya Reyes, Mario Lopez Rodriguez, Mario Humberto Mijes Cruz, Mireya Saraí García Vázquez, Jenny Benois-Pineau, Luis Miguel Zamudio Fuentes, Elizabeth Cano Martinez, Jesús Abimelek Flores Secundino, Jose Luis Rivera Martinez, et al. Image annotation for Mexican buildings database. In *SPIE Optical Engineering+ Applications*, pages 99700Y–99700Y. International Society for Optics and Photonics, 2016.

OT01. Aude Oliva and Antonio Torralba. Modeling the shape of the scene: A holistic representation of the spatial envelope. *International Journal of Computer Vision*, 42(3):145–175, 2001.

Pav77. T. Pavlidis. *Structural Pattern Recognition*. Springer-Verlag, 1977.

PBJ92. Boris Polyak and A B. Juditsky. Acceleration of stochastic approximation by averaging. 30:838–855, 07 1992. https://www.researchgate.net/publication/236736831_Acceleration_of_Stochastic_Approximation_by_Averaging.

PKL⁺12. Julien Pinquier, Svebor Karaman, Laetitia Letoupin, Patrice Guyot, Rémi Mégret, Jenny Benois-Pineau, Yann Gaëstel, and Jean-François Dartigues. Strategies for multiple feature fusion with hierarchical HMM: application to activity recognition from wearable audiovisual sensors. In *ICPR*, pages 3192–3195. IEEE Computer Society, 2012.

PMB13. Razvan Pascanu, Tomas Mikolov, and Yoshua Bengio. On the difficulty of training recurrent neural networks. In *Proceedings of the 30th International Conference on Machine Learning, ICML 2013, Atlanta, GA, USA, 16–21 June 2013* [DBL13], pages 1310–1318.

Pra91. William K. Pratt. *Digital image processing, 2nd Edition*. A Wiley-Interscience Publication. Wiley, 1991.

Pra13. William K Pratt. *Introduction to digital image processing*. CRC press, 2013.

Rab89. L. R. Rabiner. A tutorial on hidden Markov models and selected applications in speech recognition. In *Proceedings of the IEEE*, volume 77, pages 257–286, 1989.

RHGS15. Shaoqing Ren, Kaiming He, Ross B. Girshick, and Jian Sun. Faster R-CNN: towards real-time object detection with region proposal networks. *CoRR*, abs/1506.01497, 2015. http://arxiv.org/abs/1506.01497.

RHW86. David E. Rumelhart, Geoffrey E. Hinton, and Ronald J. Williams. Neurocomputing: Foundations of research. pages 696–699, 1986. http://www.nature.com/nature/journal/v323/n6088/pdf/323533a0.pdf.

Ros58. Frank Rosenblatt. The perceptron: a probabilistic model for information storage and organization in the brain. *Psychol Rev.*, 65(6):386–408, 1958.

Ros61. Frank Rosenblatt. Principles of neurodynamics. Perceptrons and the theory of brain mechanisms. Technical report, Cornell Aeronautical Lab Inc Buffalo NY, 1961.

Sha15. Gayane Shalunts. Architectural style classification of building facade towers. In *International Symposium on Visual Computing*, pages 285–294. Springer, 2015.

SHh+14. Nitish Srivastava, Geoffrey Hinton, Alex Hevsky, Ilya Sutskever, and Ruslan Salakhutdinov. Dropout: A simple way to prevent neural networks from overfitting. *Journal of Machine Learning Research*, 15:1929–1958, 2014.

SHS11. Gayane Shalunts, Yll Haxhimusa, and Robert Sablatnig. Architectural style classification of building facade windows. In *International Symposium on Visual Computing*, pages 280–289. Springer, 2011.

SHS12. Gayane Shalunts, Yll Haxhimusa, and Robert Sablatnig. Classification of gothic and baroque architectural elements. In *Systems, Signals and Image Processing (IWSSIP), 2012 19th International Conference on*, pages 316–319. IEEE, 2012.

SILH+16. Lauge Sørensen, Christian Igel, Naja Liv Hansen, Merete Osler, Martin Lauritzen, Egill Rostrup, Mads Nielsen, Alzheimer's Disease Neuroimaging Initiative, the Australian Imaging Biomarkers, and Lifestyle Flagship Study of Ageing. Early detection of Alzheimer's disease using MRI hippocampal texture. *Human brain mapping*, 37(3):1148–1161, 2016.

Sím96. Jiří Síma. Back-propagation is not efficient. *Neural Networks*, 9(6):1017–1023, 1996.

SLJ+14. Christian Szegedy, Wei Liu, Yangqing Jia, Pierre Sermanet, Scott E. Reed, Dragomir Anguelov, Dumitru Erhan, Vincent Vanhoucke, and Andrew Rabinovich. Going deeper with convolutions. *CoRR*, abs/1409.4842, 2014. http://arxiv.org/abs/1409.4842.

SMDH13a. Ilya Sutskever, James Martens, George Dahl, and Geoffrey Hinton. On the importance of initialization and momentum in deep learning. pages III–1139–III–1147, 2013. http://dl.acm.org/citation.cfm?id=3042817.3043064.

SMDH13b. Ilya Sutskever, James Martens, George E. Dahl, and Geoffrey E. Hinton. On the importance of initialization and momentum in deep learning. In *Proceedings of the 30th International Conference on Machine Learning, ICML 2013, Atlanta, GA, USA, 16–21 June 2013* [DBL13], pages 1139–1147.

SZL13. Tom Schaul, Sixin Zhang, and Yann LeCun. No more pesky learning rates. 28(3):343–351, 2013. https://arxiv.org/pdf/1206.1106.pdf.

TMLP+02. Nathalie Tzourio-Mazoyer, Brigitte Landeau, Dimitri Papathanassiou, Fabrice Crivello, Olivier Etard, Nicolas Delcroix, Bernard Mazoyer, and Marc Joliot. Automated anatomical labeling of activations in SPM using a macroscopic anatomical parcellation of the MNI MRI single-subject brain. *Neuroimage*, 15(1):273–289, 2002.

Tou74. Gonzalez R.C. Tou, J.T. *Pattern Recognition Principles*. Addison-Wesley, 1974.

UvdSGS13. J. R. R. Uijlings, K. E. A. van de Sande, T. Gevers, and A. W. M. Smeulders. Selective search for object recognition. *International Journal of Computer Vision*, 104(2):154–171, 2013. https://ivi.fnwi.uva.nl/isis/publications/2013/ UijlingsIJCV2013/UijlingsIJCV2013.pdf.

Vap92. Vladimir Vapnik. Principles of risk minimization for learning theory. In *Advances in neural information processing systems*, pages 831–838, 1992.

Vap95. Vladimir N. Vapnik. *The Nature of Statistical Learning Theory*. Springer-Verlag, Berlin, Heidelberg, 1995.

VDC12. Eleonora Vig, Michael Dorr, and David Cox. Space-variant descriptor sampling for action recognition based on saliency and eye movements. In *Computer Vision – ECCV 2012*, pages 84–97, Berlin, Heidelberg, 2012. Springer Berlin Heidelberg.

Vit67. Andrew J. Viterbi. Error bounds for convolutional codes and an asymptotically optimum decoding algorithm. *IEEE Trans. Information Theory*, 13(2):260–269, 1967.

VJ01. Paul Viola and Michael Jones. Rapid object detection using a boosted cascade of simple features. In *Computer Vision and Pattern Recognition, 2001. CVPR 2001. Proceedings of the 2001 IEEE Computer Society Conference on*, volume 1, pages I–I. IEEE, 2001.

WF12. Gezheng Wen and Li Fan. Large scale optimization - lecture 4. 2012. http://users.ece.utexas.edu/~cmcaram/EE381V_2012F/Lecture_4_ Scribe_Notes.final.pdf.

Woo02. David S Wooding. Eye movements of large populations: Ii. deriving regions of interest, coverage, and similarity using fixation maps. *Behavior Research Methods, Instruments, & Computers*, 34(4):518–528, 2002.

XTZ+14. Zhe Xu, Dacheng Tao, Ya Zhang, Junjie Wu, and Ah Chung Tsoi. Architectural style classification using multinomial latent logistic regression. In *European Conference on Computer Vision*, pages 600–615. Springer, 2014.

YTQ12. Xianfeng Yang, Ming Zhen Tan, and Anqi Qiu. CSF and brain structural imaging markers of the Alzheimer's pathological cascade. *PLoS One*, 7(12):e47406, 2012.

ZF13. Matthew D. Zeiler and Rob Fergus. Visualizing and understanding convolutional networks. *CoRR*, abs/1311.2901, 2013. http://arxiv.org/abs/1311.2901.

ZSGZ10. Bailing Zhang, Yonghua Song, Sheng-uei Guan, and Yanchun Zhang. Historic Chinese architectures image retrieval by SVM and pyramid histogram of oriented gradients features. *International Journal of Soft Computing*, 5(2):19–28, 2010.

Printed in the United States
By Bookmasters